EDUCATION IN A COMPETITIVE AND GLOBALIZING WORLD

THE TEACHER'S BOOK OF ENGINEERING THERMODYNAMICS

EDUCATION IN A COMPETITIVE AND GLOBALIZING WORLD

Additional books in this series can be found on Nova's website under the Series tab.

Additional E-books in this series can be found on Nova's website under the E-book tab.

ENGINEERING TOOLS, TECHNIQUES AND TABLES

Additional books in this series can be found on Nova's website under the Series tab.

Additional E-books in this series can be found on Nova's website under the E-book tab.

EDUCATION IN A COMPETITIVE AND GLOBALIZING WORLD

THE TEACHER'S BOOK OF ENGINEERING THERMODYNAMICS

RAFAEL NIETO CARLIER,
IGNACIO LÓPEZ PANIAGUA,
ÁNGEL JIMÉNEZ ÁLVARO,
JAVIER RODRÍGUEZ MARTÍN,
AND
CELINA GONZÁLEZ FERNÁNDEZ

Nova Science Publishers, Inc.
New York

For permission to use material from this book please contact us:
Telephone 631-231-7269; Fax 631-231-8175
Web Site: http://www.novapublishers.com

Additional color graphics may be available in the e-book version of this book.

LIBRARY OF CONGRESS CATALOGING-IN-PUBLICATION DATA

The teacher's book of engineering thermodynamics / editors, Ignacio Lspez Paniagua ... [et al.].
p. cm.
Includes index.
ISBN 978-1-61470-258-0 (hardcover)
1. Thermodynamics. I. Lspez Paniagua, Ignacio.
TJ265.T375 2011
621.402'1--dc23

2011024360

Published by Nova Science Publishers, Inc. † New York

Contents

About This Book

This book is aimed mainly for those who are teachers of Engineering Thermodynamics. It can also be very useful for the student who wants to read the subject alone; however, one should know beforehand that the text here may not be easy to follow. It is short because it is dense, so progress could be slow in pages and fast in concepts. For the student, it is adviceable to have in hand friendlier sources of information, but to use this book as a guide.

The structure of the information contained in the book has been made explicit to the reader, to provide the option to choose what to read and what to leave out as a safe criterion. If a chapter is conceptually diff cult, a map of the main ideas involved and some relations between them is available at the beginning, to allow the reader to keep track of the conceptual coherence of it all. The major conceptual milestones are outlined. This map should also serve as a reminder of what should be kept in mind after processing the chapter.

We have also considered it useful to insert some comments in the text. Generally on conceptual issues that students usually don't think about at all, but that may be of interest for the teacher, as well as some hints about what kind of things students usually f nd hard to understand from our own classroom experiences. So, the text within this book is organized with def nitions and Laws highlighted, some useful exercises, and is interrupted here and there by boxed comments for the teacher; like in the following example:

DEFINITION: A def nition explains the meaning of a word or expression. This is the following def nition.

Law 6 *This would be the Sixth Law of Thermodynamics, if such thing happened to exist.*

Behavior in the limit -1 *There is no knowledge about the behavior of the matter in the limit of $P \to \infty$*

Don't forget that listening to classes shouldn't be boring.

PROOF

As a general rule, we shall *f rst* enunciate important statements, concepts or ideas and *then* give a proof. This is the opposite order to the normal texts, in which the conclusion is shown as a f nal consequence of a reasoning. However, our version is clearer and more useful to the teacher and to the mature student, since there is a more immediate access to how the ideas, however complex, (most of them result of lengthy reasonings) relate to each other.

Proofs may sometimes be rather long and we want the reader to know exactly where the Proofs begin and where they end; so this has been made clear within the text. Boxes could perhaps be misleading, as these comments are remarks indicated for the teacher. We shall use lines, resembling the upper-left and the lower-right corners of a box, like here.

As the subject is 'Engineering Thermodynamics,' we have kept the engineer in mind throughout. We have tried to cover the main concepts and applications of interest to an engineer. In short, engineers use Thermodynamics mainly for three purposes:

1. To make a balance of the energy in a process (First Law)

2. To study the quality of the energy involved in a process (Second Law, best as balance of exergy)

3. To calculate thermodynamic properties, needed for these balances and also in other applications.

So, for example, exergy plays a signif cant role in this book, as it is a quite useful tool in engineering; on the other hand, such intriguing matters as inf nities and scalation laws near the critical point, in spite of being very interesting, are not devoted a single word in this text, as engineers will hardly ever have to deal with them.

We have assumed that the approach used by the teacher here, is one that uses Classical Thermodynamics. This is mainly for two reasons: f rst, that it is used by many Universities; and second, because in this way only simple backgrounds in Mathematics and Physics are needed. We have found it impossible to make an approach to Thermodynamics without at least requiring previous knowledge about these subjects; since it would imply signif cant losses of generality and rigour. In spite our goal for the most direct and less demanding way to present this subject, we have included a chapter on Statistical Thermodynamics, giving a simple introduction to it, to help connect the macroscopic and the microscopic points of view.

So, topics studied in this book are:

- Basic concepts and Zeroth, First and Second Laws (chapters 1 to 3).

- Thermodynamic potentials, equilibrium and stability (chapter 4).

- Short introduction to Statistical Thermodynamics, including Third Law (chapter 5).

- Single component phase equilibrium and thermodynamic properties calculation (chapters 6 and 7).

- Open system balances (chapter 8).

- Thermodynamic property calculations for multicomponent systems (chapter 9).

- Thermodynamic balances and chemical equilibrium for reacting systems (chapter 10).
- Brief study of selected engineering applications of Thermodynamics (chapter 11).

Throughout the book, we have tried to project a clear distinction between the Laws of Thermodynamics and some issues about the behavior of substances at one or another limit conditions, which we think has not been stressed enough in the literature. The Laws, usually called Zeroth, First and Second, each introduced a new function of state: T, U, and S. We regard this role at a higher degree of importance than that of what we have called the *conditions of behavior in the limit*, namely, (a) *in the limit of null pressure*, related with the model of an ideal gas an a mixture of ideal gases; (b) *in the limit of zero temperature*, that it is usually called 'Third Law'; and (c) *in the limit of 1 and 0 mole fraction* related to the ideal mixture model and the model of Henry, respectively. These conditions of limit do not introduce new functions of state, and 'only' tell how real matter behaves in certain limit situations. So, in our approach, there are seven basic independent points, where each is a generalization of many experiences:

1. Basic postulate of Thermodynamics (chapter 1, page 3)
2. Zeroth Law (chapter 1, page 6)
3. First Law (chapters 2 and 8)
4. Second Law (chapters 3 and 8)
5. Behavior in the limit $P \to 0$ (chapter 2, for pure substances, page 10 and 28; and for multicomponent systems, chapter 9 page 131)
6. Behavior in the limit $T \to 0$ (Third Law, section 5.6.)
7. Behavior in the limit $x \to 1$ (Lewis-Randall ideal mixture, section 9.7.) and $x \to 0$ (Henry model, section 9.9.)

Unfortunately, Thermodynamics uses a lot of different functions, and there is no general agreement about notation. Different authors use different notations and, what it is worse, none of them are completely consistent. So, a consistent notation, if the case were possible, would be rather different from the usual practices. We have tried to be consistent, but not to differ too much from what can be found used by other authors.

As a general rule, we have represented extensive functions by capital letters (except in the case of mass, m, and quantity of substance, n, that have no specifc functions), and the corresponding specifc function by the same letter but lowercase. Intensive non-specifc functions are special cases: we have used P for pressure and T for temperature, as this is almost universally accepted (and t has been used for empiric temperature).

We have tried to name different things with different symbols. As an example of the kind of troubles we have had, remember that the Helmholtz function is denoted in some books by F and in others by A; none of these options is without drawbacks, as fugacity is everywhere represented by f (that should correspond to the specifc Helmholtz function with the f rst option) and activity is represented by a (that should correspond to the specif c Helmholtz function in the other case). By the way, our choice was a for the specif c Helmholtz function and $\mathfrak{a}$ for activity.

To give another example, the symbol Δ is everywhere used as an operator with the meaning $\Delta Z = Z_2 - Z_1$, where Z is any function, and 2 and 1 represent two given states. Unfortunately, all books (well, not all; our own book is an exception) on Chemical Thermodynamics use the same symbol for the partial function of reaction (which is in fact a derivative); and, combined in the form ΔZ°, for the standard function of reaction (which is a sum). Any choice has its drawbacks, as trying to be consistent will make the notation completely different from all other books notation. We have chosen a very similar but nevertheless different symbol, $\triangle Z^{\circ}$. In this way, we have managed to use a different symbol for every different meaning, but without looking too different from all the other books.

Chapter 1

Temperature and Thermal Equation of State

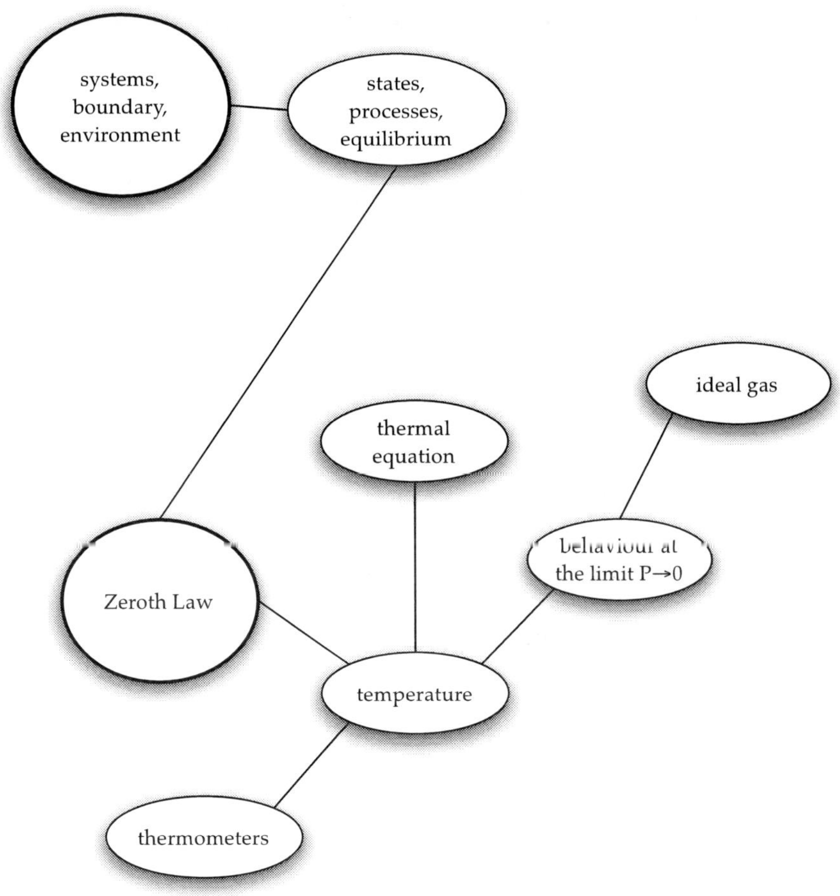

In this chapter, we shall prepare the most basic conceptual starting points of Thermodynamics. First of all, and what may be most frequently ignored by students, we shall define *what* Thermodynamics is about: *systems*. We shall see how to describe systems in a thermodynamically meaningful way –state variables– which allows not only to characterize how the system *is* at a certain instant, but to calculate interesting variables when it eventually changes for some reason.

These notions, system, environment, boundary, state variables, processes, and equilibrium, will drive us to the Zeroth Law, which will hint one of the most fundamental concepts in Thermodynamics: *temperature*. We shall then spend some time in thinking about temperature and how to compare this property among systems through *temperature scales*. This will lead us to an amazing characteristic of nature: the *ideal gas temperature*, a f rst glimpse of the behavior of matter at the limit $P \to 0$, which will leave our deliberations about temperature one step before conclusion; the end will arrive in chapter 3 with the concept of *thermodynamic temperature*. In relation with all these ideas, we shall end by def ning what is normally understood as an *ideal gas* which, as we shall never remark enough, is a *model*, and *does not* stand for gases at low pressures.

1.1. Basic Concepts

1.1.1. System, Environment, Boundary

DEFINITION: The system is the part of the universe that we want to study.

DEFINITION: The environment is the part of the universe external to the system, that interacts appreciatively with it.

DEFINITION: The boundary is the frontier between the system and its environment.

DEFINITION: The *universe* is the system together with its surroundings.

- What exactly is the system, is just a matter of convenience.
- The separation between the interior and the exterior of a system forms its *boundary*. It can be conceptual or physical, and the researcher must decide on whether to count it as part of the system or not.
- The environment is what is not considered as part the system. It means the zone that interacts with the system, not *all* the rest of the Universe.
- System and environment are mutually related: whatever happens within the system triggers effects in its exterior and vice versa: they interact.

If the boundary is impermeable, there is no mass interchange between the system and its environment, and the system is said to be *closed*. Otherwise, it is *open*.

1.1.2. State and Equilibrium

DEFINITION: State is the set of values of all measurable properties of a system, at a given instant.

A system that does not interchange mass (closed) nor energy with its environment is *isolated*.

Table 1.1. Classif cation of boundaries.

Criterion	Type	Description
Permeability	Impermeable (wall)	The boundary prevents mass interchange between the system and its environment.
	Semipermeable (membrane)	The boundary allows mass interchange between system and environment, but only of some components.
	Permeable	The boundary allows mass interchange of any component between system and environment.
Volume	Rigid	The volume of the system may not change (neither in value or shape).
	Deformable	The volume of the system may change.
Energetic interaction across its boundary	Adiabatic	The system may only interchange energy with its environment in the form of work.
	Diathermal	The system may interchange heat with its environment.

Make a clear distinction between *isolated* and *thermically isolated.*

DEFINITION: A state is of equilibrium if, keeping the system isolated, the measurable properties of the system do not change in time.

Basic Postulate of Thermodynamics: Any isolated system spontaneously reaches a unique state of equilibrium.

This postulate serve to get rid of some highly idealized systems that may cause conceptual diff culties, like an ideal pendulum. Such systems may oscillate without never reaching equilibrium, thus making reversible cycles through non-equilibrium states. To avoid considering this surprising behavior, which real systems never present, we just exclude these systems from consideration.

We shall describe the state of a system by means of a set of variables which represent its properties. For example: volume, pressure at a given point, length, mass, and so on. According to the previous def nition, the state of the system is the set of values of these variables at a given instant.

In order to describe the state of a system in equilibrium, it is not necessary to consider *all* of its inf nite properties, but a minimal number which we shall call *degrees of freedom* of the system. Knowing the values of these, the value of the rest can be deduced. The variables which we choose to def ne the state of equilibrium of a system are called *state variables.* They must be mutually independent, and they must be equal in number to the degrees of freedom of the system. They can be freely chosen as long as these conditions are satisf ed.

Mathematical formulation: We shall call X_i, $i = 1 \dots l$ the state variables, l the degrees of freedom in the system, and Z will indicate a generic function.

We may know the value of a variable Z as a function of the state variables, so that

$$Z = Z(X_i)$$

We may observe that, as any state variable can be transformed into a different variable by means of a function, state variables may be interchanged as long as the resulting set remains mutually independent.

1.1.3. Processes

DEFINITION: Given a certain system S at state 1, we call process to the evolution of the system and the interaction with its environment when the system evolves to another state, 2.

Equilibrium Thermodynamics can only study processes in which at least the initial and f nal states are of equilibrium.

DEFINITION: A special type of process in which all intermediate states are of equilibrium is called quasistatic process.

This happens whenever the velocity at which the process takes place is suff ciently slow as to allow the system to reach the equilibrium at any intermediate instant.

When this does not happen, the system may deviate appreciatively from equilibrium during the process. Then, the process is called *non-static.*

Sometimes it is possible to def ne and formulate non-equilibrium states by considering *internal variables* in addition to the state variables that describe the equilibrium states. In this case, the processes in which these states (and only these in addition to equilibrium states) intervene can be treated mathematically as if they were quasistatic, only involving more independent variables.

DEFINITION: A process is *reversible* whenever it can be inverted, that is, when an *inverse process* exists by which both system and environment may evolve simultaneously from the f nal to the initial state.

It must be observed that only quasistatic processes might be reversible, but that a quasistatic process might not be, necessarily, reversible.

Description of a Process in Terms of State Variables

We shall indicate by Z a generic state function of the system, and the change of state by ΔZ. A generic process will be indicated by $\mathcal{P}$. In general, the initial and f nal states of a process will not coincide, leading to a change of state for the system when the process ends: $\mathcal{P} \Rightarrow \Delta Z$.

During the process, the system travels across a series of states $\mathcal{R}$. In the case of a quasistatic process, these states are equilibrium states, and the series of intermediate states is called *state trajectory* of the process. For example, in a normal system (no metric variable except for volume), there are two degrees of freedom, the thermodynamic space of equilibrium is a plane (like P-v) and the line of process is a line in this plane; the sequence of states for a non-static process is not drawable because the intermediate states are not points on the plane of equilibrium.

In order to fully describe the process, it is necessary to determine the interaction between the system and its environment, which is specifc to each process. In fact, even for a same $\mathcal{R}$, the interaction between the system and the environment could be different in each process. The magnitudes associated to the interaction of system and environment are called *line functions*.

The general description of a process is:

$$\mathcal{P} \Leftrightarrow (\mathcal{L}_i, \mathcal{L}_i^e) \Rightarrow \mathcal{R} \Rightarrow \Delta Z$$

$(\mathcal{L}_i, \mathcal{L}_i^e)$ indicate the line variables of the system and of the environment respectively. If a system is in equilibrium, a process will start only because the system would receive an excitation from its environment, through the boundary. Mathematically, this means that a system-environment interaction takes place, measured from the environment by the line variables of the environment, $\mathcal{L}_i^e$, and from the system by its corresponding line variables, $\mathcal{L}_i$. A series of intermediate states, $\mathcal{R}$, takes place through these variables, and, as a consequence, a change of state ΔZ.

> $(\mathcal{L}_i, \mathcal{L}_i^e)$: as an example, a work given by the system is received by the environment, so they have a different sign. Even the absolute value may be different if there is a surface effect at the boundary.

Mathematical properties: Z–state function. $\mathcal{L}$–line function.

$dZ = M\,dx + N\,dy$	$d\mathcal{L} = M\,dx + N\,dy$
$\left(\frac{\partial M}{\partial y}\right)_x = \left(\frac{\partial N}{\partial x}\right)_y$	$\left(\frac{\partial M}{\partial y}\right)_x \neq \left(\frac{\partial N}{\partial x}\right)_y$
$\oint dZ = 0$	$\oint d\mathcal{L} = \mathcal{L} \neq 0$
The integral is path independent	The integral is path dependent

1.1.4. Extensive and Intensive Variables

Consider a system at a certain equilibrium state and another system, identical to the former, at the same state. Evidently, the mass of the set of both systems is double to that of a single system of the two. In these conditions, the variables which have the same value when computed for the set than when computed for the initial system are called *intensive* y_i (examples: pressure, temperature, etc.). The variables whose value has doubled when passing from the original system to the set, and are therefore proportional to the total mass, are called *extensive* X_i, and hold the mathematical relation of **degree 1** homogeneity:

$$\lambda X = X(\lambda m_i, \lambda X_i, y_i) \tag{1.1}$$

with m_i being the masses of the different components of a system. We shall represent extensive variables with capital letters.

The ratio of extensive variables by unit of mass or quantity of matter are called *specif c variables*.

$$x = \frac{X}{m} \text{ (1 component)} \qquad x = \frac{X}{\sum_{i=1}^{C} m_i} \text{ (}C\text{ components)} \tag{1.2}$$

being X an extensive variable proportional to mass, specif c variables must be intensive; for example, specif c volume $v = V/m$. We shall represent the specif c variables with the same letter as the corresponding extensive one, in lowercase.

Another type of variable that is sometimes used is the *density of function*

$$\check{x} = \frac{X}{V} \tag{1.3}$$

which is the value of an extensive function by unit of volume (for example, the well known mass density $\rho = m/V$). These type of variables are frequently used when studying electric or magnetic systems.

It follows from the def nition that every extensive variable is also *additive*, which means that its value for the total system is the addition of its value for each of the parts.

In moving systems, *external variables* must be considered in addition to the state variables, which describe the interactions due to the presence of f elds (mechanical, electrical or magnetic) and the movement of the system (linear velocity, angular velocity, position). In many cases –though not always– they are decoupled from the state variables.

> DEFINITION: A system is *homogeneous* when all its specif c variables have the same value in all points of the system.

A non-homogeneous system is called *heterogeneous*

> DEFINITION: A *phase* is an homogeneous subsystem of an heterogeneous system.

1.2. Zeroth Law and Temperature

> DEFINITION: Thermal equilibrium is the equilibrium reached by two systems put in contact through a diathermal wall.

It must be remembered that diathermal walls allow interchanging energy in a form other than work (that is, they allow heat interchange). The existence of thermal equilibrium will imply a relation between the state properties of each system. If we consider two systems in thermal equilibrium, 1 and 2, with their states def ned by two variables x,y (for simplicity); there will be a relation of the kind:

$$F(x^2, y^2, x^1, y^1) = 0$$

.

Law 0 *Two systems that are in thermal equilibrium with a third system are in thermal equilibrium with one another.*

Let us develop this a little further. Let us imagine a certain system 1, in thermal equilibrium with a certain system 2, which is, in turn, in thermal equilibrium with a third system 3, and let us imagine that 1 and 3 are separated by an adiabatic wall. Then, in spite of the adiabatic wall, it is known that systems 1 and 3 would also be in thermal equilibrium if the wall were to be diathermal.

As we shall see immediately, this implies the existence of a new function of state, called temperature.

> In fact, there is not just one temperature, but many. Different thermometers measure different empirical temperatures. Later, we shall f nish the study of temperature with two temperatures that are independent of thermometer: ideal gas temperature (in this chapter) and thermodynamic temperature, that will appear with the Second Law.

This law allows us to build thermometers (which perform the role of system 2). By means of a thermometer, we may measure the temperature of a certain system and we know that any other system that were at the same temperature, even if it were miles away, would be in thermal equilibrium with the f rst.

1.2.1. Thermometers

Consider systems S1, S2, and S3, in equilibrium states given by their state variables (two each for simplicity):

$$S1 \equiv (x^1, y^1),\ S2 \equiv (x^2, y^2),\ S3 \equiv (x^3, y^3)$$

When the systems are in thermal equilibrium, their state variables are related. We can formulate this relation mathematically with a function. Therefore, if S2 is in thermal equilibrium with S1 and S3, there will exist two functions such that:

$$F_A = F_A(x^2, y^2, x^1, y^1) = 0$$
$$F_B = F_B(x^2, y^2, x^3, y^3) = 0$$

We can work out x^2 in both:

$$x^2 = f_A(y^2, x^1, y^1) = f_B(y^2, x^3, y^3)$$

In order for system 2 to be useful as a thermometer, it is designed so that y^2 either remains constant or is a function of x^2: the thermometer is chosen so that it has a single degree of freedom. In this way, y^2 disappears from the previous equation and we can describe the state of equilibrium of systems 1 and 3 only with x^2:

$$x^2 = f_A(x^1, y^1) = f_B(x^3, y^3)$$

This expression shows that, according to the Zeroth Law, systems 1 and 3 have a common state variable that we shall call *temperature*.

In industry, the most widely spread thermometers are thermocouples, in which the independent variable is electric potential. In liquid column thermometers the independent variable is the length of the column.

Constant-volume gas thermometers work in an analogous way. A thermometer is a deposit with a pressure gauge, plus some devices to ensure constant volume (compensating for the variations produced by the gauge). The deposit is closed and rigid, and in it, a certain mass of gas is conf ned. The independent variable is the pressure inside the deposit, $x^2 = P$.

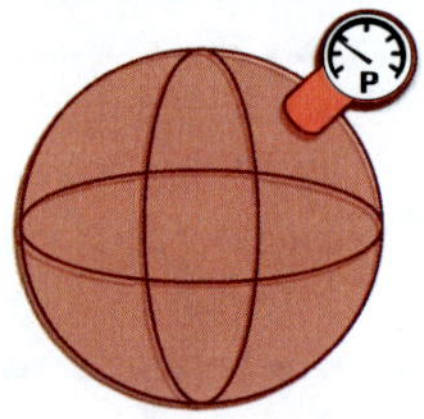

It must be observed that gas thermometers are also empirical, that means the value of temperature which they measure depends on the characteristics of the device itself. In particular, for a gas thermometer, the relevant characteristics are the nature and density of the gas conf ned in its deposit.

Another possibility would be a gas thermometer at constant pressure (a cylinder with a non-friction piston). It is not used.

The variable x^2 is never used directly as temperature. It is called *thermometric variable*, but the *temperature* is a function of it:

$$t = t(x^2)$$

Therefore, the value of temperature which is actually measured depends on the actual thermometer used (through x^2) and on the form of the function (*thermometric scale*). Given that the function is arbitrary, a simple choice is generally made. For a gas thermometer, $t = t(P) = K \cdot P$, with K a constant for the thermometer and P the pressure in the deposit.

1.2.2. Temperature Scales

Historically, the Celsius scale was def ned on two points:

- Point of fusion of ice: $0\,^{\circ}\mathrm{C}$
- Boiling point of liquid water: $100\,^{\circ}\mathrm{C}$

Both were taken at standard ambient pressure (101 325 Pa). Logically, given that the real ambient pressure may vary depending on altitude, weather, etc., the def nition of the scale was subject to signif cant experimental errors.

Currently, the temperature scale is def ned taking the following into account:

- The triple point of water, which may only occur at f xed conditions of temperature and pressure, regardless the geographical location.

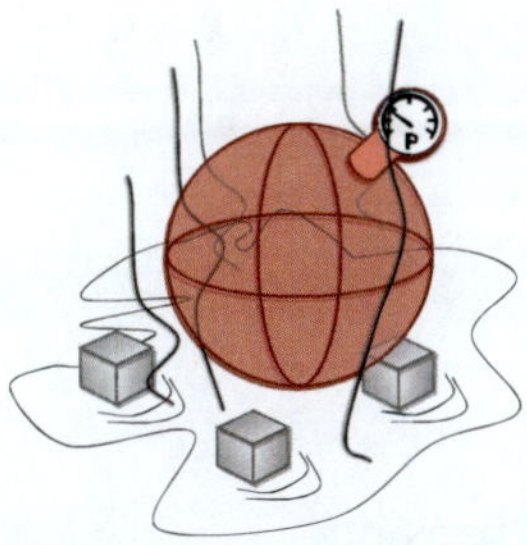

Triple point: solid/liquid/gas in equilibrium (611 Pa, 0.01 °C). Thermometric deposit in red.

- The value of temperature at the triple point must be 0.01 °C for coherence with the old def nition of the scale.

Now, let us consider that we have a certain quantity of water at its triple point, in equilibrium with a gas thermometer. The pressure in the deposit will have a certain value, P_{T_t}. The Kelvin temperature scale for a gas thermometer will be def ned as:

$$t = 273.16\,\mathrm{K} \cdot \frac{P}{P_{T_t}} \tag{1.4}$$

and the Celsius by:

$$t(^{\circ}\mathrm{C}) = t - 273.15^{\circ}\mathrm{C} \tag{1.5}$$

Remark that different thermometers built like this would in fact measure different values, because the gas in each thermometer (a real gas) would have its own, characteristic, properties: nature and density. In this way, the student gets prepared for the concept of ideal gas.

In spite of the fact that from now on until the last section of the chapter, we are going to use T to indicate *ideal gas temperature*, this notation should really indicate *thermodynamic temperature*. Conceptually, both variables are different. However, we shall see further on that their values coincide.

1.2.3. Thermal Equation of an Homogeneous System

DEFINITION: We shall call thermal state equation of a system to the relation between its volume (either extensive or specif c), pressure and temperature, such as $v = v(P,T)$ or $P = P(T,V)$.

DEFINITION: The *compressibility factor* of a substance is $z = \frac{PV}{nRT}$

with $R = 8.314472$ J/mol K, n number of moles y T the ideal gas temperature. The thermal equation is frequently expressed as $z = z(P,T)$.

The way to measure the ideal gas temperature, that appears in this equation and the following equations, will be seen in this chapter, section 1.3.1.

Deriving the thermal equation $v = v(T,P)$ and dividing by v:

$$\frac{dv}{v} = \frac{1}{v}\left(\frac{\partial v}{\partial T}\right)_P dT + \frac{1}{v}\left(\frac{\partial v}{\partial P}\right)_T dP$$

The thermal coeff cients are def ned:

$$\boxed{\alpha = \frac{1}{v}\left(\frac{\partial v}{\partial T}\right)_P} \quad \text{Coeff cient of volume expansion} \tag{1.6}$$

$$\boxed{\kappa = -\frac{1}{v}\left(\frac{\partial v}{\partial P}\right)_T > 0} \quad \text{Isothermal compressibility coeff cient} \tag{1.7}$$

$$\boxed{\beta = \frac{1}{P}\left(\frac{\partial P}{\partial T}\right)_v} \quad \text{Coeff cient of tension} \tag{1.8}$$

so that: $\frac{dv}{v} = \alpha dT - \kappa dP$.

> The sign in the def nition of κ ensures that always $\kappa > 0$. This is a consequence of stability, that lies outside the scope of this book. It must be mentioned, however, that the sign of α and β can be either positive or negative, as long as they both have the same. It is normally positive but, in some substances and under particular conditions it may turn negative. This, for example, is the case of water at ambient pressure and temperatures lower than 4 °C (maximum density).

Given that v, P, and T are state variables, they have perfect differentials, and therefore satisfy:

$$\left(\frac{\partial \alpha}{\partial P}\right)_T = -\left(\frac{\partial \kappa}{\partial T}\right)_P$$

Thus, remembering the mathematical property:

$$\left(\frac{\partial x}{\partial y}\right)_z \left(\frac{\partial z}{\partial x}\right)_y \left(\frac{\partial y}{\partial z}\right)_x = -1$$

and applying it to P, v and T, we have:

$$\alpha = P\kappa\beta \tag{1.9}$$

1.3. Behavior in the Limit of Null Pressure

Different gases share common characteristics in their behavior at very low pressures; it is interesting to observe this pattern, which we can express in terms of V, P, T as:

> **Behavior in the limit 1** *(1st part) For any pure gas:*
>
> $$\lim_{P\to 0} \frac{PV}{nRT} = \boxed{\lim_{P\to 0} z = 1} = z^* \qquad \textit{where } R = 8.314472\,\text{J/mol K.}$$

The apparent exceptions are due to systems that are really not a pure gas, but react to form dimers or decompose.

T in this expression is the *ideal gas temperature*. In the next subsection we shall study this concept, which is inspired precisely on the fact that different gases have a common pattern of behavior at low pressures. We shall also study the ideal gas model, for which we shall reserve a specif c notation, given the importance it will have in future chapters: its properties at given T and P shall be indicated by a superscript $*$ (for example, v^* will be the specif c volume of the model).

1.3.1. Ideal Gas Temperature

DEFINITION: The ideal gas temperature of a system at a given state is given by:

$$T = \lim_{P_{T_t} \to 0} t = \lim_{P_{T_t} \to 0} 273.16\text{K} \cdot \frac{P}{P_{T_t}}$$

where t is the empirical temperature measured by the gas thermometer and, as it was already mentioned when describing the gas thermometer, P is the pressure of the gas in the constant volume gas thermometer in thermal equilibrium with the system whose ideal gas temperature we want to measure, and P_{T_t} is the pressure in the thermometer when it is in equilibrium with water at its triple point.

Near the limit at null pressure, the density with which we start to measure is irrelevant; so the single specif c characteristic of the thermometer which could affect the measurement would be the nature of the gas itself.[1] But, as we shall see in the following, the ideal gas temperature def ned thus is also independent from the gas itself, a fact which may be deduced from the Gay-Lussac law.

In summary, the *ideal gas temperature* is independent both from the thermometer with which it is measured and from the gas inside. This implies that, unlike with many other variables, there exists an *absolute* temperature with which to compare others. The justif cation that follows will help understanding –and believing– this.

PROOF

Consider a gas thermometer holding a certain quantity of gas 1, which measures the temperature of a certain system at a certain, f xed state. The thermometer must have been calibrated by putting it in thermal equilibrium with water at triple point and measuring the pressure in the deposit $P_{T_t,1}$. We shall measure the temperature of the desired system by putting thermometer and system in thermal equilibrium and reading the pressure in the deposit, which will be P_1. The actual value of empirical temperature will result from the following formula

$$t_1^1 = 273.16\,\text{K}\frac{P_1}{P_{T_t,1}}$$

and it will be the (empirical) temperature of the system measured with the gas 1-thermometer with density 1. If we now extract part of the original gas contained in the

[1] Alternatively, we could express this idea in terms of the limit at null density instead of at null pressure.

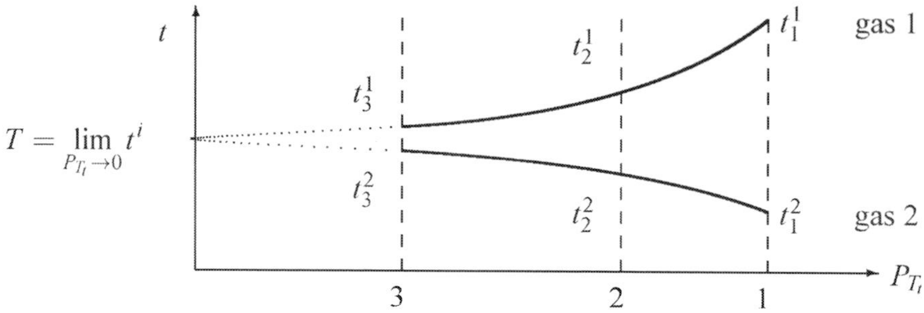

Figure 1.1. Ideal gas temperature.

thermometer, we shall have a new, **different**, gas thermometer, because we shall have altered the density. Calibrating the new thermometer by putting it in contact with water at triple point, we shall obtain a new, lower value of its pressure $P_{T_t,2}$. Measuring the temperature of the system with this thermometer (gas 1, density 2) by reading a new value of pressure P_2, will lead to the empirical temperature

$$t_2^1 = 273.16\,\text{K}\frac{P_2}{P_{T_t,2}}$$

somewhat different to t_1^1, because the thermometer is different. We may continue extracting gas from the deposit, achieving lower and lower densities in the deposit, calibrating and reading each time lower pressures when measuring the temperature of the system, until we eventually reach the limit at null density, or equivalently $P_{T_t} \to 0$.

Repeating the process with another gas 2 in the thermometer, we would obtain another series of values for the empirical temperatures that we would measure, as shown in f gure 1.1. For example, given that in both cases the thermometers would be calibrated against the same reference (water at triple point), the Gay-Lussac law which we shall see in page 13, implies that the limit at null pressure will coincide for both gases and for any other that we could chose. The limit is therefore *independent from the thermometer.*

There exists an International Practical Scale of Temperatures that is used as a reference. The ideal gas temperatures of the f xed points of the scale are obtained at specialized laboratories by costly experiments with gas thermometers.
The measurement error of commercial thermometers is referred to this scale.

1.3.2. Empirical Laws at Low Pressures

In the following, t is the temperature measured with a constant volume gas thermometer.

Boyle law: It holds:

$$\lim_{P \to 0} \frac{PV}{m} = f_i(t) \neq 0$$

with $f_i(t)$ depending only of the temperature and the nature of the gas i.

Gay-Lussac law: Consider a gas i in two given states, occupying the same volume in both: (P_1, t_1) and (P_2, t_2). Then, if P_1 and P_2 are sufficiently low, it holds:

$$\lim_{P_1 \to 0} \frac{P_2}{P_1} = \psi(t_2 - t_1) \qquad \forall i$$

being $\psi(t_2 - t_1)$ the same function for any gas i which we would consider, that is, independent from the thermometer. It may be remarked that, taking $t_1 = 273.16\,\text{K}$, this implies that the limit mentioned in section 1.3.1. is the same for any gas.

Avogadro law: In a given volume, at equal pressure and temperature at the limit of null pressure, the number of moles of any gas is the same.

These empirical laws are the experimental basis from which the behavior of the thermal equation of state at the limit of null pressure was deduced.

1.4. Ideal Gas

DEFINITION: The ideal gas is the *mathematical model* defined by the thermal equation

$$z^* = \frac{PV^*}{nRT} = 1 \Rightarrow \boxed{PV^* = nRT} \Rightarrow Pv^* = RT$$

where the asterisk indicates the value given by the model (not of a real system).

It has been verified empirically that the behavior of the model resembles the behavior of real gases at low pressures; z tends to 1, which is its value for the ideal gas model, and some other properties (but not all) behave similarly.

Chapter 2

First Law

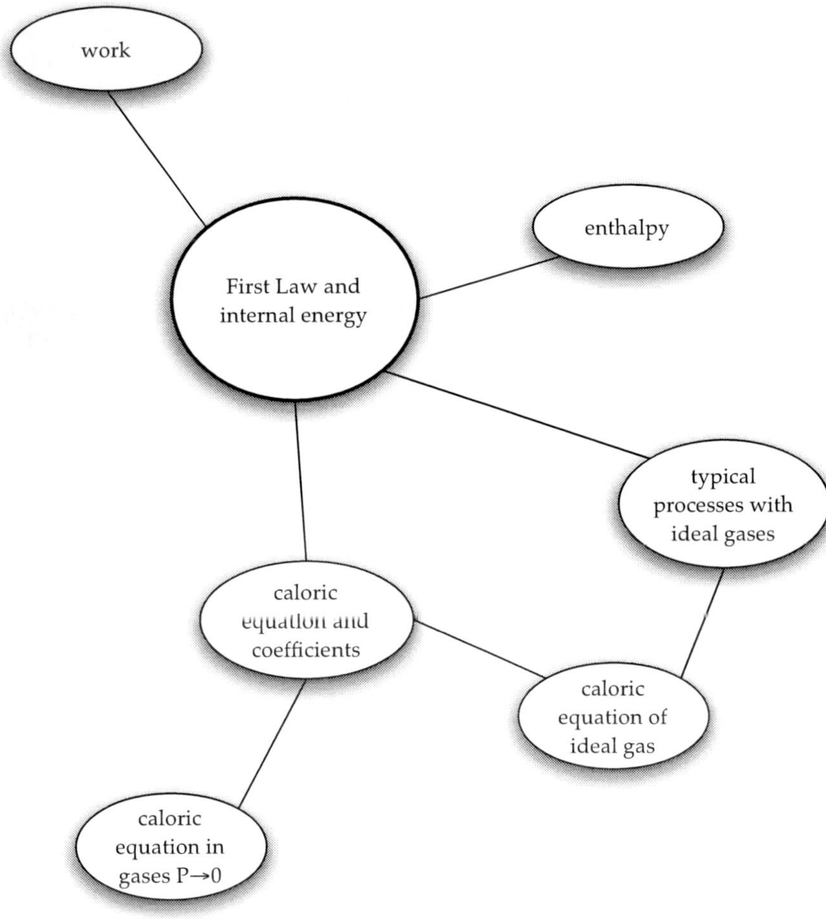

Historically, people did not know that boiling water in a kettle and lifting a weight were related by an abstract property we now call *total energy*, sum of kinetic, potential, and internal energy contributions, that we commonly use. An intuition that one could have to do with the other was built very early, but it was not until relatively recently that actual numbers and values could be put to heat, work, and energy in the same expression, and that they were related to our most common working variables: pressure, temperature, and volume.

Even today, the Second Law is generally regarded as the most abstract by the students, but our long experience in the classroom shows that most students have misconceptions of internal energy and, more surprisingly, of work. Indeed, once the equivalence between work and heat was established, it seemed diff cult to distinguish them until Gibbs' def nition arrived. More, calculating work in a process means def ning perfectly what the system to be considered will be and what the process will be like, and both require having understood the most general notion of all: system. The problem lies in the fact that, in a system through a process, heat, work and energy, may appear disguised in many ways: with falling weights, chemical reactions, magnetic f elds, mixtures and many others. It then may become confusing to distinguish the system with boundary and forces.

In this chapter, we shall revise work in its most general form, which holds for any type of force and system. Then, we visit the common type found in thermodynamics: work due to variation of volume, but the reader should bear in mind that the First Law holds for *all* types of work, only that we use this as an useful example. After the concept of work is established, we introduce the concept of internal energy, trying to stress its physical meaning just to bring it out of the abstract as much as we can. Then, we shall deal with some issues of direct interest to the engineer: how is it formulated (caloric equation) in general and how this is particularized for the simple case of an ideal gas. This will allow us to complete the portrait of the behavior of substances at the null pressure limit. On the way, we shall have introduced the concept of *enthalpy*, alternative to that of internal energy, which in practice turns out to be of much more frequent use to the engineer.

2.1. Generalized Work

> DEFINITION: Work is performed by a system on its surroundings during a given process if the *only* effect external to the system could be the raising of a weight.

Remark «only».

The def nition is for work performed by the system on its environment (positive); exchanging the roles of system and environment, the work performed by the environment on the system can also be considered (negative).

In general, work can be expressed as:

$$\boxed{dW = y_{ex} dX} \tag{2.1}$$

with y_{ex} an intensive variable of the surroundings called generalized force, and dX an extensive variable of the system called generalized displacement.

The previous def nition of work indicates that if, given a process, we could couple a device to the original system so that the only external result of the process would be the raising of a weight, we can say that the system and its environment were interchanging energy as *work*. However, if this were not possible, and any possible device that could be coupled to the system would necessarily produce other effects apart from the raising of a weight, then the system and the environment did not interact energetically only by work.

Fundamental criterion: Thermodynamics only considers the work that the system interchanges with its environment. That is, the work which results from agents and forces external to the system. The internal dynamics of the system *does not* perform work in the thermodynamic sense of the term. In other words: forces whose point of application falls within the boundaries of the system and move internal parts, subsystems that may eventually change their relative location *within the system*, etc. *do not* perform work.

2.1.1. Mechanical Work Associated to Volume Variation

In general, the mechanical work that a force performs is defned by $dW = \vec{F} \cdot d\vec{l}$, being $d\vec{l}$ the differential displacement of its point of application. According to the fundamental criterion already mentioned, the mechanical work in Thermodynamics considers forces *external* to the system, $\vec{F}_e$. More, the sign convention adopted requires work to be positive when performed by the system, so:

$$dW = -\vec{F}_e \cdot d\vec{l}$$

Let us see how this formulation applies to the case of a system under a certain exterior pressure. The force on a certain surface element of the system ($d\vec{A}$) is: $d\vec{F}_e = -P_e d\vec{A}$. If, during a process, the element is displaced by a differential, $d\vec{l}$, the work performed by the system is:

$$dW = -d\vec{F}_e \cdot d\vec{l} = P_e d\vec{A} \cdot d\vec{l} = P_e dV$$

We may observe that dV is the differential volume which is swept by the element of surface along its displacement.

Therefore, the work associated to a variation of volume of the system during a process between instants i and f is:

$$\boxed{W = \int_i^f P_e dV} \tag{2.2}$$

However, this integral may be solved only by knowing the variation of P_e with the volume of the system: $P_e = P_e(V)$. This is not straightforward in general. Sometimes, P_e can be expressed as a function of the internal pressure of the system, related to its volume by its thermal equation and/or restrictions of different kinds. Nevertheless, this is not always possible; in some processes the internal pressure of the system is not defned.

2.1.2. Other Kinds of Work

Apart from the work associated to the variation of the system volume, there exist other types of work. For example, a system can also interact electrically and magnetically with its environment, interchange work through a rotating shaft, etc. Of course, there may also exist a mechanical work of dissipation whose force would be tangent to the system boundary.

For example, electric work could be formulated as $dW = \mathcal{V}\, d\mathcal{Q}$, where $d\mathcal{Q}$ would represent the electric charge interchanged between the system and its environment, and $\mathcal{V}$ the voltage at which the interchange would take place.

2.1.3. Example of Calculating the Work Associated to Variation of Volume

We are going to solve a problem in which the work performed by the system is analyzed for different cases.

Remark the choosing of the system-environment boundary

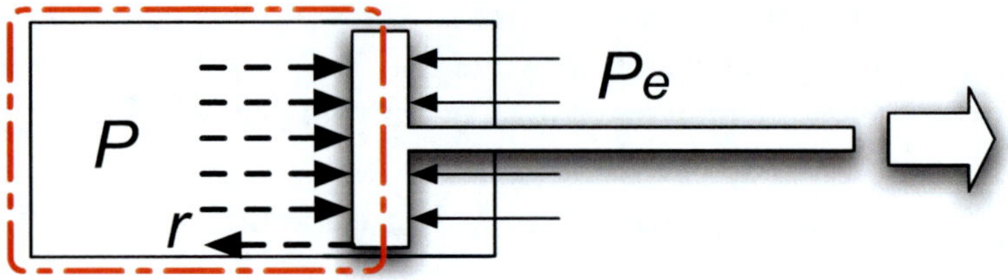

Exercise 1: Consider a system formed by a cylinder of diathermal walls and piston, containing n moles of an ideal gas. If the system is sunk in a thermostatic bath at constant temperature T_1, its pressure becomes P_1. Calculate the work performed by the system on its environment, on the following cases:

1. The gas expands at negligible speed and without friction until its pressure equals P_2.
2. A constant friction force of value Ar exists between the piston and the cylinder, being A the cross section of the cylinder. The gas expands at negligible pressure until P_2.
3. The only pressure external to the system is the atmospheric pressure P_0. There is no friction, and the gas expands until mechanical equilibrium is reached.
4. The opposing forces to the expansion of the gas are the atmospheric pressure and the friction of case 2. The piston stops at $P_4 = P_0 + r$.

Solution: We choose the system so that it includes the piston. In this way, the friction, eventually, will be an internal force and will not be counted for calculating the work.

1. Quasistatic, non-friction process:
   - There exists only work associated to variation of volume: $W = \int_1^2 P_e\, dV$
   - Mechanical equilibrium: $P_e = P$; ideal gas $PV = nRT_1 \quad \Rightarrow$

$$W = \int_1^2 P_e\, dV = \int_1^2 P\, dV = \int_1^2 \frac{nRT_1}{V} dV = \boxed{nRT_1 \ln \frac{P_1}{P_2}}$$

2. Quasistatic process with friction:
   - There exists only work associated to variation of volume: $W = \int_1^2 P_e\, dV$
   - Mechanical equilibrium: $P_e = P - r \quad \Rightarrow$

$$W = \int_1^2 P_e\, dV = \int_1^2 (P - r) dV = \boxed{nRT_1 \left[\ln \frac{P_1}{P_2} - r\left(\frac{1}{P_2} - \frac{1}{P_1}\right)\right]}$$

If P_2 were equal to the f nal pressure of the previous case, the work would be smaller.

3. In this case, $P_e = P_0$. It is a non-static process, therefore internal pressure would not even be def ned.

$$W = \int_1^3 P_e dV = P_0 \int_1^3 dV = P_0 \Delta V = nRT_1 \left(1 - \frac{P_0}{P_1}\right)$$

4. The process is analogous to the previous one, except for the addition of friction. The f nal pressure at the cylinder is higher than in the previous case and the f nal volume smaller.

$$W = \int_1^4 P_e dV = P_0 \int_1^4 dV = P_0 \Delta V = P_0 nRT_1 \left(\frac{1}{P_0 + r} - \frac{1}{P_1}\right)$$

The work performed by the system is smaller than in the previous case, because the piston reaches equilibrium earlier.

Remark that if the same f nal volume were f xed by mechanical means in the last two cases, the work performed by the system would be *the same* in both cases.

2.2. First Law and Internal Energy

Law 1 *Consider a closed system at rest in which there exist no relativistic effects. There exists a state function called* internal energy *whose increment is equal to the difference between the heat it receives from its environment and the work it performs on it:*

$$\boxed{\Delta U = Q - W} \tag{2.3}$$

In moving systems, the total energy *must be considered instead:*

$$\boxed{\Delta U' = \Delta U + \Delta E_c + \Delta E_p = Q - W} \tag{2.4}$$

In engineering, nuclear reactions are the most frequently encountered relativistic effect. The First Law does not hold for nuclear reactions, because energy is not conserved; a more general principle of conservation of mass/energy should then be used.

PROOF

It can be empirically demonstrated that a resting system with no relativistic effects, when following two adiabatic processes from state 1 to state 2, performs the same work in both cases, whichever these two processes. As this is a characteristic property of state functions, we can conclude that system work in adiabatic processes is equal to the increment of a state function:

$$W^A = -\Delta U$$

The negative sign is conventional. In resting, non-relativistic closed systems following non-adiabatic processes, the heat absorbed by the system is def ned by:

$$Q = \Delta U + W$$

which is therefore another than work form of interchanging energy between systems.

The internal energy of a system is a macroscopic concept which includes all the types of energy that could exist within a system. Figure 2.1 shows some of the types that may exist.

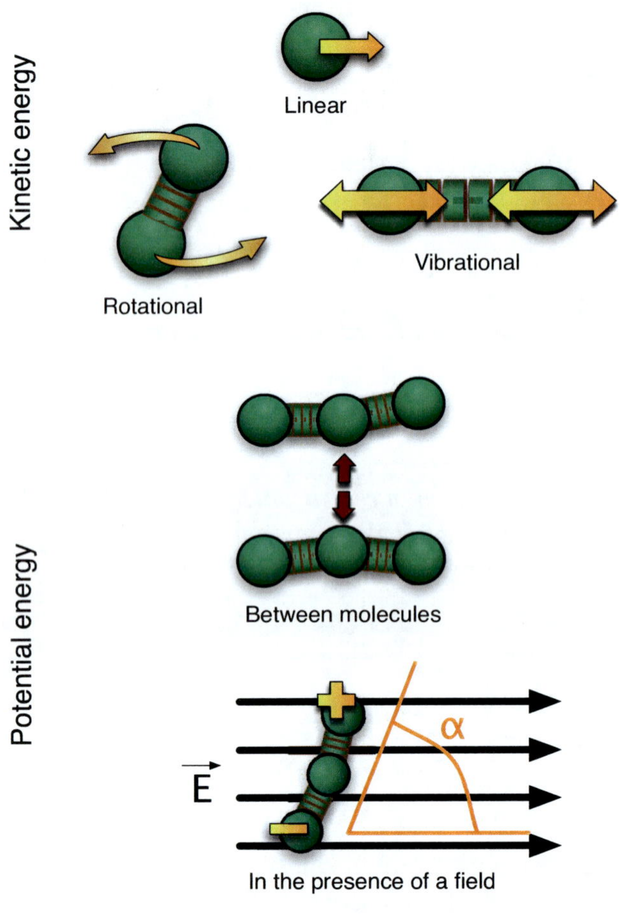

Figure 2.1. Examples of different modes of storing internal energy.

2.2.1. Properties of the Internal Energy

- Internal energy is a state function (U) while work (W) and heat (Q) are line functions. They all have energy units and will be measured in J in the International System of Units.
- Internal energy is additive. If a system is formed by several subsystems (i, $i =$

$1,\dots,n$), each with internal energy U_i measured with compatible references, then:

$$U = \sum_i U_i$$

- It follows from the additivity of internal energy:

$$\frac{U}{m} = u$$

for an homogeneous system, being m the total mass of the system, and u the *specific internal energy*.

It is frequently advisable to calculate with specific values and convert the result to extensive ones at the end. This is why we shall use specific values preferably from now on.

It must be remarked that internal energy as such is not defined. Only its *increment* is; therefore, its origin, $U = 0$, is arbitrary. Hence, internal energies of two different systems are not comparable except when one system can be transformed into the other maintaining it closed and without relativistic effects throughout the process (as by a chemical reaction).
The increment of internal energy is always defined for a closed system, regardless whether it is chemically reactive or whether it is multi-component (a mixture). This implies that compatible references must be adopted for different chemical species, because some could be transformed into others by a chemical reaction. As we just mentioned, the energy balance of the First Law would hold provided that the system would remain closed. We shall return to this point in chapter 10.

In the cases of an adiabatic process and a process without work, it holds, respectively:

$$\Delta U = -W^A \qquad \Delta U = Q$$

which means that the work performed by the system in the first case comes from decreasing its internal energy, and that the heat received by the system in the second case increases its internal energy.

In both cases, we may observe that a line function (W^A or Q) coincides with a state function (U). They are two specific cases which only occur under the mentioned circumstances.

2.2.2. Examples: First Law in Moving Systems

Exercise 2: We consider the cylinder of figure 2.2, with diathermal walls. The length of the spring when no force act upon it equals the length of the cylinder, and its constant is $k = 4\,\text{N/m}$. The thickness of the piston and its friction with the cylinder walls are negligible. Through the valve indicated in the figure, a given amount of an ideal gas, with $c_v = 1.5R$ is introduced, thus compressing the spring to a length $l_0 = 0.5\,\text{m}$.

Heat is slowly given to the gas, thus expanding it and performing work on the spring. Find the heat absorbed by the gas when its volume doubled the initial volume.

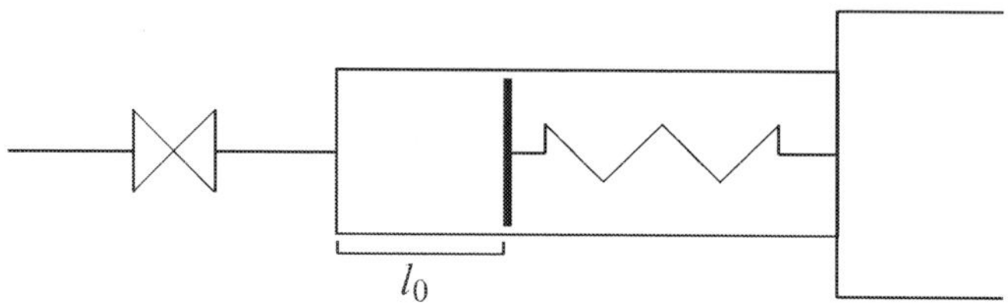

Figure 2.2.

Solution: Calling A to the cross section of the cylinder, l the length the spring has been compressed from its initial state to a given state,

$$V = l\mathsf{A} \quad ; \quad P = \frac{kl}{\mathsf{A}} \Longrightarrow PV = kl^2 = nRT \Rightarrow T = \frac{kl^2}{nR}$$

As there is no friction, the work is

$$dW = PdV = \frac{kl}{\mathsf{A}}\mathsf{A}dl = kl\,dl \Rightarrow W = \int_{l_0}^{2l_0} kl\,dl = \frac{3}{2}kl_0^2$$

As we shall see in this chapter, section 2.4.3., the caloric equation of state for an ideal gas of constant c_v is

$$\Delta U = nc_v(T - T_0) = \frac{c_v}{R}(4kl_0^2 - kl_0^2) = \frac{3c_v kl_0^2}{R}$$

$$Q = \Delta U + W = 3kl_0^2\left(\frac{1}{2} + \frac{c_v}{R}\right) = 3 \times 4\frac{\mathrm{N}}{\mathrm{m}}\,0.25\,\mathrm{m}^2\left(\frac{1}{2} + 1.5\right) = 6\,\mathrm{J}$$

Exercise 3: A rigid and solid object is propelled upwards, and stops at an height of 50 km with a temperature $t = 50\,°\mathrm{C}$. It falls to the ground, reaching a velocity $c_2 = 600\,\mathrm{m/s}$. Find its temperature in this moment, knowing that 40% of the work of friction is disipated to the atmosphere. Assume that there is no work of deformation of the solid.

Solution: Kinetic and potential energy in initial and f nal states:

$$e_{p1} = \mathsf{g}z_1 = 9.8\frac{\mathrm{m}}{\mathrm{s}^2}\,50\,\mathrm{km} = 490\,\mathrm{kJ/kg} \quad ; \quad e_{c1} = 0$$

$$e_{c2} = \frac{1}{2}c_2^2 = \frac{1}{2}(600\,\mathrm{m/s})^2 = 180\,\mathrm{kJ/kg} \quad ; \quad e_{p2} = 0$$

We shall consider the system formed by the solid **excluding its surface**, so that the frictional force is external to the system and its work must be computed in the First Law (negative sign). We have been told that 40% of this work f ow as heat through the surface to the atmosphere, so the remaining 60% should f ow as heat **inwards** (positive sign); this is the heat that should be counted in the First Law according with our choice for the def nition of the system. From Mechanics, it is well known that the work of the frictional force is equal to the variation of mechanical energy, so

$$\Delta u + \Delta e_c + \Delta e_p = q - w = -0.6(\Delta e_c + \Delta e_p) - (-\Delta e_c - \Delta e_p) \Rightarrow$$
$$\Rightarrow \Delta u = -0.6(\Delta e_c + \Delta e_p) = 186\,\text{kJ/kg}$$
$$du = c\,dt + \left(\frac{\partial u}{\partial v}\right)_T dv = c\,dt \Rightarrow \Delta u = c\,\Delta t$$
$$\Delta t = \frac{186\,\text{kJ/kg}}{0.25\,\text{kcal/kg K}} = 177.8\,\text{K} \Rightarrow t_2 = t + \Delta t = 227.8\,^\circ\text{C}$$

Alternatively, we could have considered as system the solid **including** its surface and even the part of the air where the friction takes place; in that case, the work would be zero, and the heat would also be different. The reader should write the First Law with this choice and verify that the temperature obtained is the same (as it neccesarily should be, because it is a measurable variable and cannot depend on our choice of system).

2.3. Enthalpy

DEFINITION: Enthalpy is a state function def ned by:

$$\boxed{H = U + PV} \tag{2.5}$$

Enthalpy has units of energy. Analogously to internal energy, it is additive and the specif c enthalpy is: $h = \frac{H}{m}$.

It must be remarked that, as both enthalpy and internal energy are related, the origin of one of them is immediately def ned when setting an arbitrary origin for the other. Both origins, however, are not the same. For example, taking $U(P_0, V_0) = 0$ we have $H(P_0, V_0) = P_0 V_0 \neq 0$.

> As will be shown in chapters 8 through 11, enthalpy is generally more useful than internal energy for an engineer, because enthalpy plays a major role in the energy balance of open systems.

2.3.1. Measuring Internal Energy and Enthalpy

As we have seen, in an adiabatic process: $\Delta U = -W^A$. This means that the variation of internal energy can be measured by measuring work in adiabatic processes. Analogously, in processes without work, $\Delta U = Q$, all the heat received by the system increases its internal energy.

In the case of enthalpy, if we consider a process in which the initial and f nal pressures are equal to the exterior pressure, it follows:

$$\Delta H = \Delta U + \Delta(PV) = Q - \cancel{P_e \Delta V} + \cancel{P_i \Delta V} = Q$$

The heat received by the system is equal to the increment of its enthalpy (only in this type of processes). This is especially useful when analyzing chemical reactions.

2.4. Caloric Equations

DEFINITION: We call *caloric equation* of a system to any of the expressions:

$$U = U(T,V) \quad ; \quad H = H(T,P)$$

In order to consider equations describing U or H to be caloric equations, it is necessary that the independent variables be the ones of the previous def nition.

2.4.1. Caloric Coeff cients

Both U and H are state functions which we consider differentiable and continuous for an homogeneous system. Instead of working with extensive variables, we shall take their intensive ones:

$$u = \left(\frac{\partial u}{\partial T}\right)_v dT + \left(\frac{\partial u}{\partial v}\right)_T dv = c_v dT + l_\theta dv$$

$$h = \left(\frac{\partial h}{\partial T}\right)_P dT + \left(\frac{\partial h}{\partial P}\right)_T dP = c_p dT + \lambda_\theta dP$$

c_v	specif c heat at constant volume	l_θ	energetic coeff cient of isothermal expansion
C_v	heat capacity at constant volume: $C_v = mc_v$		
c_p	specif c heat at constant pressure	λ_θ	enthalpic coeff cient of isothermal compressibility
C_p	heat capacity at constant pressure: $C_p = mc_p$		

2.4.2. Properties of the Caloric Coeff cients

Physical sense: Specif c heats stand for the increment of energy/enthalpy which results from increasing the temperature of the system by one unit, under constant volume/pressure. They are state functions.

Derivatives: As u and v are state functions, the following holds:

$$\left(\frac{\partial c_v}{\partial v}\right)_T = \left(\frac{\partial l_\theta}{\partial T}\right)_v$$

$$\left(\frac{\partial c_p}{\partial P}\right)_T = \left(\frac{\partial \lambda_\theta}{\partial T}\right)_P$$

Relation between caloric coeff cients: Given that u and v are related, the caloric coeff-cients of one can be expressed in terms of the other:

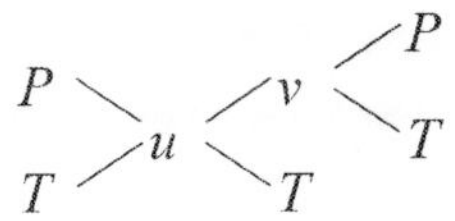

therefore

$$c_p = \left(\frac{\partial h}{\partial T}\right)_P = \left(\frac{\partial u}{\partial T}\right)_P + P\left(\frac{\partial v}{\partial T}\right)_P = c_v + l_\theta \alpha v + P\alpha v = c_v + \alpha v(l_\theta + P) \qquad (2.6)$$

$$\lambda_\theta = \left(\frac{\partial h}{\partial P}\right)_T = \left(\frac{\partial u}{\partial P}\right)_T + v + P\left(\frac{\partial v}{\partial P}\right)_T =$$

$$= v + \left(\frac{\partial v}{\partial P}\right)_T (l_\theta + P) = v[1 - \kappa(P + l_\theta)] \Rightarrow \boxed{l_\theta = \frac{1 - \frac{\lambda_\theta}{v}}{\kappa} - P} \qquad (2.7)$$

As it will be demonstrated in page 49, l_θ (or, conversely, λ_θ) and $c_p - c_v$ can be deduced from the thermal equation.

2.4.3. Caloric Equation of the Ideal Gas

The following holds for an ideal gas:

$$u^* = u^*(T)$$
$$h^* = h^*(T)$$

The reason for this is that $\left(\frac{\partial u^*}{\partial v^*}\right)_T = T\left(\frac{\partial s^*}{\partial v^*}\right)_T - P = \frac{RT}{v^*} - P = 0$, as will be proven in equation (3.11).

The asterisk will indicate that the property refers to an ideal gas. The specif c internal energy of an ideal gas only depends on temperature; and the same applies for its enthalpy, because $h = u + Pv$, and for an ideal gas $Pv^* = RT$.

We shall now see two properties of ideal gases which are related by their caloric equation:

Caloric coeff cients of an ideal gas:

$$c_v^* = \left(\frac{\partial u^*(T)}{\partial T}\right)_v = \frac{du^*}{dT} = c_v^*(T) \qquad l_\theta^* = \left(\frac{\partial u^*(T)}{\partial v}\right)_T = 0$$

$$c_p^* = \left(\frac{\partial h^*(T)}{\partial T}\right)_P = \frac{dh^*}{dT} = c_p^*(T) \qquad \lambda_\theta^* = \left(\frac{\partial h^*(T)}{\partial P}\right)_T = 0$$

Relation between c_p^* and c_v^* (Mayer's relation):

$$c_p^* = \left(\frac{\partial[u^*(T) + Pv^*]}{\partial T}\right)_P = \frac{du^*}{dT} + R = c_v^* + R$$

2.5. First Law and Typical Processes with Ideal Gases

Given that $l^*_\theta = 0$ and $\lambda^*_\theta = 0$ for ideal gases, it always holds:

$$PV = nRT$$
$$\Delta U = \int_i^f nc_v dT$$
$$\Delta H = \int_i^f nc_p dT$$

We are going to apply these equations to some typical processes with ideal gases; in particular, processes throughout which V, P or T remain constant, $Q = 0$ or in which $PV^k =$ constant. Except when indicated otherwise, we shall assume that there only exists work due to variation of volume.

2.5.1. Constant Volume Process: $V =$ const.

The process may be quasistatic or non-static, because the extensive volume of the system is always def ned (oppositely, the specif c volume might not, for in a non-static process it may have different values at different parts within the system). The work of variation of volume is:

$$W = \int_i^f P_e dV = 0 \Rightarrow \boxed{\Delta U = Q} \Rightarrow \Delta H = \Delta U + \Delta(PV) = \Delta U + V\Delta P$$

2.5.2. Constant Pressure Process: $P =$ const.

The process must be quasistatic in order to guarantee the system pressure be def ned throughout. The work of variation of volume is:

$$W = \int_i^f P_e dV = P_e \Delta V \Rightarrow \Delta U = Q - P_e \Delta V$$

In addition, if it occurred $P = P_e$, as it was mentioned when studying enthalpy:

$$\Delta H = \Delta U + P\Delta V = Q - \cancel{P\Delta V} + \cancel{P\Delta V} = Q$$

It must be remarked that this expression would be valid even in a non-static process, as long as the initial and f nal pressures of the system are equal to the external pressure, and this is constant.

2.5.3. Constant Temperature Process: $T =$ const.

The process must be quasistatic to guarantee that the system temperature is def ned throughout.

$$\Delta H = \int_i^f nc_p dT = 0 \quad ; \quad \Delta U = \int_i^f nc_v dT = 0 \Rightarrow Q = W$$

It must be observed that the previous expressions hold even for a non-static process, provided $T_i = T_f$ and the substance is an ideal gas. In the specifc case of a non-frictional, quasistatic process, $P = P_e$

$$W = \int_i^f P_e dV = \int_i^f P dV = nRT_1 \ln \frac{V_2}{V_1} = -nRT_1 \ln \frac{P_2}{P_1}$$

2.5.4. Adiabatic Process: $Q = 0$

The process may be quasistatic or non-static. The state trajectory is not fully determined by the process being adiabatic; there may exist alternative ones or it might not even be defned.

$$W = -\Delta U = -\int_i^f nc_v dT$$

For an example of how the state trajectory of the process could be determined, when possible, refer to section 2.5.5.

2.5.5. Example: *The only equation for adiabatic processes is* $Q = 0$

We are going to analyze the behavior of an ideal gas with constant $\gamma = \frac{c_p}{c_v}$. It will expand quasistatically in a cylinder, with constant friction, equivalent to a pressure of value r.

- There is force equilibrium because the process is quasistatic: $P_e = P - r$
- First Law: $dU = -dW \Rightarrow nc_v dT = -P_e dV = -(P-r)dV$
- Thermal equation: $PV = nRT \Rightarrow dT = \frac{VdP+PdV}{nR}$
- Grouping: $\frac{c_v}{R} VdP + (P\frac{c_p}{R} - r)dV = 0$
- Finally, integrating: $\left(\frac{V_1}{V_2}\right)^\gamma = \frac{P_2\frac{c_p}{R}-r}{P_1\frac{c_p}{R}-r} \Rightarrow \boxed{\left(P - \frac{rR}{c_p}\right) \cdot V^\gamma = \text{const.}}$

If, in addition to all the previous equations, the process had no friction, $r = 0 \Rightarrow PV^\gamma = const$. This equation will be most probably known by the majority of students from previous courses, but, due to the large number of *necessary* conditions for it to hold, it will seldom be useful in practice. Moreover, even in the cases in which it could be used, it could be replaced by an expression of the Second Law, gaining in simplicity and ease of operation.

2.6. Polytropic Processes: $PV^k = \text{const.}$

The process must be quasistatic, in order for the pressure of the system to be defned throughout. It may be any fuid. The state trajectory of the process is the one defning the polytropic process itself: $PV^k = const$. Some examples of polytropic processes are:

- $k = 0 \Rightarrow P = \text{const.} \Rightarrow$ Constant pressure process

- $k = 1 \Rightarrow T = \text{const.} \Rightarrow$ Constant temperature process (if ideal gas)
- $k \to \infty \Rightarrow V = \text{const.} \Rightarrow$ Constant volume process

If, in addition, the process were non-friction:

$$W = \int_i^f P\,dV = \int_i^f \frac{P_1 V_1^k}{V^k} dV = P_1 V_1^k \int_i^f \frac{dV}{V^k} =$$

$$= \frac{P_1 V_1^k}{1-k}[V_2^{1-k} - V_1^{1-k}] = \frac{P_1 V_1}{1-k}\left[\left(\frac{V_1}{V_2}\right)^{k-1} - 1\right]$$

except if $k = 1$, which results in the expression seen for the constant temperature process.

2.7. Caloric Equation in Gases at Null Pressure

Behavior in the limit 1 *(2nd part): Caloric equation in the limit of null pressure*

$$\lim_{P \to 0} \lambda_\theta = f(T) \neq 0 = \lambda_\theta^* \quad ; \quad \lim_{P \to 0} l_\theta = 0 = l_\theta^*$$

It is frequent to think that a real gas at low pressure behaves exactly like the ideal gas model. This is an error, in so many aspects, but they differ in some properties; for example, λ_θ. We are going to see how the caloric equation of a gas behaves at the null pressure limit:

Many textbooks present at this point the experience of Joule. The fact is that it failed to prove anything, as Kelvin noticed. The actual behavior in the limit was obtained by the Joule-Thomson experience. By the way, the reasoning is interesting, because is quite similar to what it will be done to pass from closed systems to open systems.

The diagram in f gure 2.3 represents an experimental setting for carrying out an adiabatic, quasistatic process with no friction at the pistons.

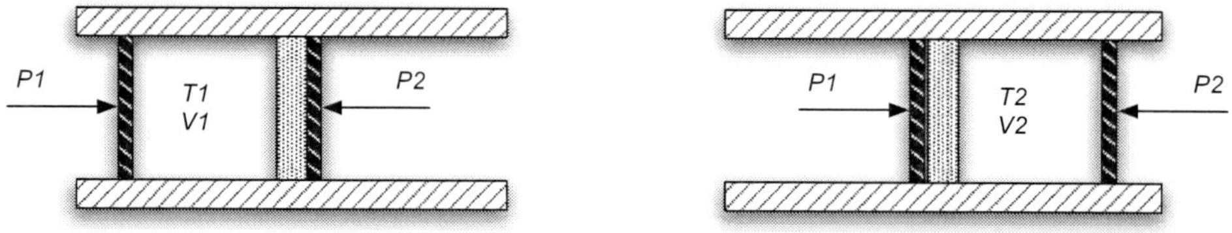

Figure 2.3. Setting for the Joule-Thomson experiment.

It consists in a cylinder with pistons at both ends. In between there is a porous wall and the whole apparatus is thermally insulated.

1. Initially, both pistons are situated at the same end of their stroke, as indicated in the f gure, so that all the gas is between one of them and the wall, in conditions P_1, T_1, V_1.
2. At the end of the process, when the pistons have moved to the opposite end, the gas is at P_2, T_2 and occupies a volume V_2, different to the initial one.

3. The exterior pressure applied on each piston is constant throughout. Applying the First Law to the set (adiabatic), it follows that it is a constant enthalpy process:

$$U_2 - U_1 = -(W_1^A + W_2^A) = P_1 V_1 - P_2 V_2 \Rightarrow$$
$$U_1 + P_1 V_1 = U_2 + P_2 V_2 \Rightarrow H_1 = H_2$$

4. By measuring the temperature difference between the initial and final states, the Joule-Thomson coefficient can be worked out:

$$\mu_{JT} = \left(\frac{\partial T}{\partial P}\right)_h = -\frac{\left(\frac{\partial h}{\partial P}\right)_T}{\left(\frac{\partial h}{\partial T}\right)_P} = -\frac{\lambda_\theta}{c_p} \approx \frac{T_2 - T_1}{P_2 - P_1}$$

by repeating the experience with lower pressures each time, it is observed that:

$$\lim_{P \to 0} \lambda_\theta = -\lim_{P \to 0} \mu_{JT}\, c_p = f(T) \neq 0$$
$$\lim_{P \to 0} l_\theta = -\lim_{P \to 0} \left(\frac{1 - \frac{\lambda_\theta}{v}}{\kappa} - P\right) c_p = 0$$

because, as a consequence of the limit behavior of the thermal equation, in these conditions $v \to \infty$, $1/\kappa \to P$ and, as we have just seen, λ_θ tends to a finite value.

In the ideal gas model, $\mu_{JT}^* = 0$ and $\lambda_\theta^* = 0$, which differ from the respective null pressure limit values for a real gas. These are some of the differences between the ideal gas model and the behavior of real gases.

Differences Ideal Gas/Real gas at null pressure.

$$\lambda_\theta^* = 0 \neq \lim_{P \to 0} \lambda_\theta \Rightarrow \mu_{JT}^* = 0 \neq \lim_{P \to 0} \mu_{JT}$$
$$v^* - \frac{RT}{P} = 0 \neq \lim_{P \to 0} \left(v - \frac{RT}{P}\right)$$

It can be easily seen which results coincide at the null pressure limit and which do not, bearing in mind that, as proven by Statistical Thermodynamics, from the null pressure limit up to moderate pressures the thermal equation of any gas has the form $z = 1 + \mathsf{B}(T)P + \mathsf{C}(T)P^2 + \ldots$, depending B, C... exclusively on temperature.

Chapter 3

Second Law

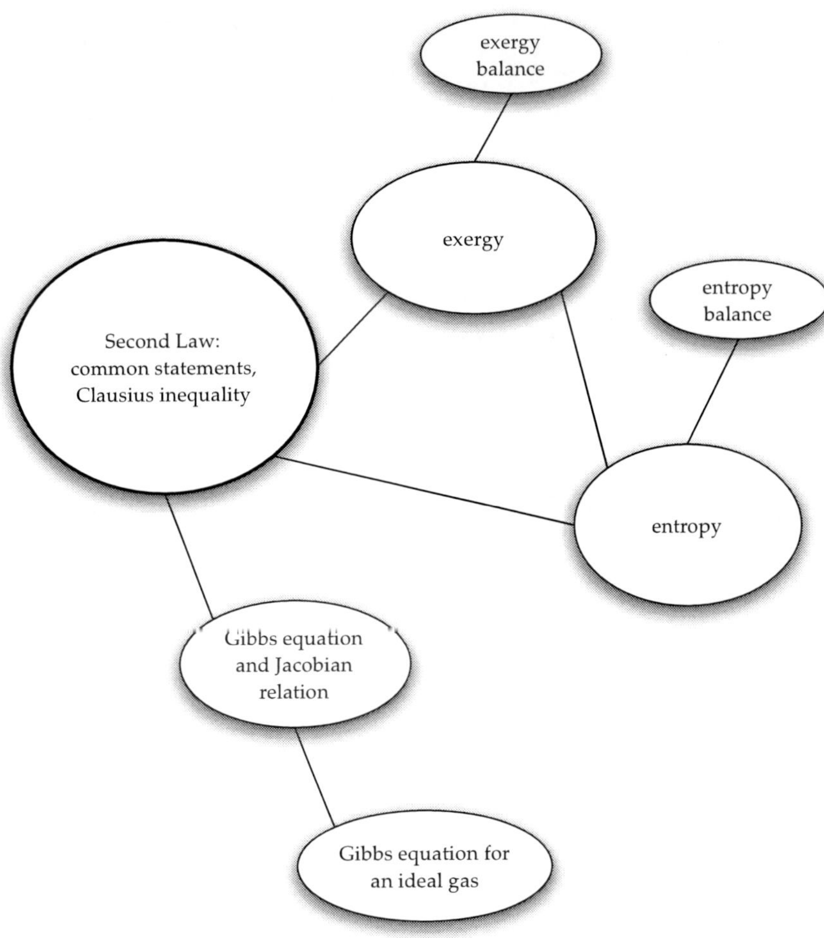

According to the First Law, heat and work could be exchanged; one for the other, without ever losing any amount of either in the process. But we all intuit this is not true and in fact, it is not. This is the Second Law. However, expressing it in a technical way with all the necessary conceptual and mathematical infrastructure to allow calculations was not easy; although, it f nally happened in the form of *entropy*, *entropy f ux* and *entropic generation*. In this chapter, we shall cover all these concepts and explain the reasoning that justif es many of the relations that exist among them. We shall f nally conclude the issue of temperature, started in chapter 1, with the def nition of the *thermodynamic temperature*.

There are several aspects of interest regarding this chapter from the point of view of the engineer. First, there is a clear difference in philosophy between the f rst and second halves of the chapter. The f rst half assesses the nature of entropy and its properties, and therefore treats entropy in itself. The second half covers entropy balances, which answer the question 'where does entropy come from –and where does it go– in a given process?', which is exactly the question of practicing engineers.

It must also be said that there are two alternative formulations of the Second Law: one based on entropy, the other on exergy. Entropy is the classical one, perhaps nearest to the origin of the concept. Exergy is more practical, because it assesses heat and work and the useful parts of both after considering irreversibilities. It is nearer to the engineer. We have paid attention to both for the conceptual insight that the former provides, but an engineer should surely feel more comfortable with the exergy balance.

3.1. Purpose of the Second Law

- To valuate the quality of the energy of a system, understanding by this its capacity to be transformed into another type of energy or the capacity of the system to produce work.
- To know the possible direction of the evolution of a process and to def ne from this the criteria for equilibrium and stability of a system.
- To valuate quantitatively the irreversibility of a process.

3.2. Entropy and Second Law

DEFINITION: The state function *entropy* is def ned by:

$$dS = \frac{dQ^R}{T} \tag{3.1}$$

where R indicates that it is the heat interchange in a reversible process, and T is thermodynamic temperature.

In order to prove that S is a state function, we shall f rst have to study the concept of thermodynamic temperature and prove the Clausius equality.

Law 2 *In any closed system with no relativistic effects, the entropy variation between two equilibrium states is equal to the sum of the entropy f ux and the entropy generated in the process ($\sigma \geq 0$).*

$$\boxed{\Delta S = J_s + \sigma \qquad (\sigma \geq 0)} \tag{3.2}$$

where the entropy f ux is $J_s = \int \frac{dQ}{T}$

The justif cation for this is conceptually demanding and lengthy, so we have split it into the following sections, ending in the actual core notion: the Clausius equality and inequality.

3.3. Previous Def nitions

Source: Closed system in which temperature, volume and any other metric variable remain constant even if the system exchanges heat.

Engine: Closed system that follows a *cyclic* process.

The engine is called *direct* if $W > 0$ and *reverse* otherwise.

3.4. Common Statements of the Second Law

Clausius (1850): It is impossible to transfer heat from a body of lower temperature to other body of higher temperature without any other effect.

Kelvin-Planck (1851/1886): It is impossible any cyclic process with no other effect than taking heat from a source and producing an equal quantity of work.

Caratheodory (1909): In any arbitrarily short interval around any state in a given thermodynamic space, there are states that are adiabatically inaccessible.

Sears-Kestin (1966): The internal energy of any closed system in any adiabatic and isometric process cannot decrease.

The f gure shows three illustrations of the postulates from Clausius, Kelvin, and Caratheodory. The f rst two cannot occur.

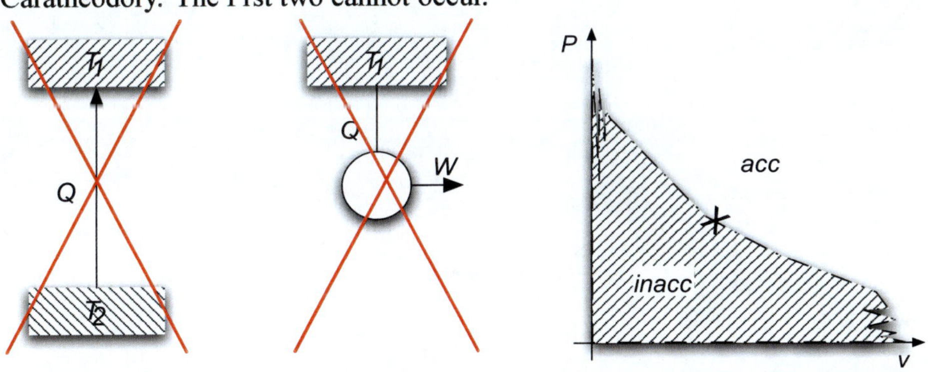

3.5. Thermodynamic Temperature

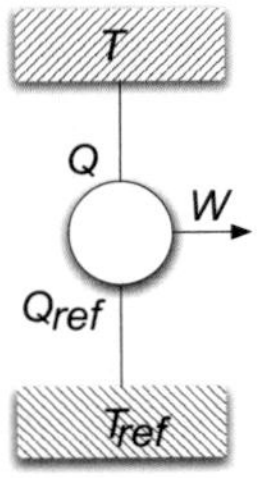

DEFINITION: The thermodynamic temperature is the value of T which satisf es:

$$\frac{T}{T_{ref}} = -\frac{Q}{Q_{ref}}$$

where $T_{ref} = 273.16\,\mathrm{K}$, and Q and Q_{ref} are the amounts of heat exchanged by a Carnot engine operating between two sources, respectively at T and T_{ref} at the temperature of the triple point of water.

Thermodynamic temperature is needed because an ideal gas-thermometer can only measure within a range between minimum and maximum values. Below the minimum, the gas would change phase, and above the maximum, the thermometer would break. This means that the ideal gas temperature would not be def ned for all possible equilibrium states.

We are going to prove that there exist a temperature –thermodynamic temperature– which is independent from the thermometer and valid for any possible equilibrium state. Later, we shall see that this temperature coincides with the ideal gas temperature whenever the latter is def ned.

For this purpose, we shall have to make some previous considerations regarding engines, Carnot cycles and Carnot's theorem.

3.5.1. Considerations on Bi-thermal Engines and Second Law

Let us consider a direct, bithermal engine, $W > 0$, as the one shown in f gure 3.1. We shall name the sources so that $T_1 > T_2 > 0\,\mathrm{K}$, and we shall indicate the amounts of heat absorbed by the engine from the sources by Q_1 and Q_2 respectively.

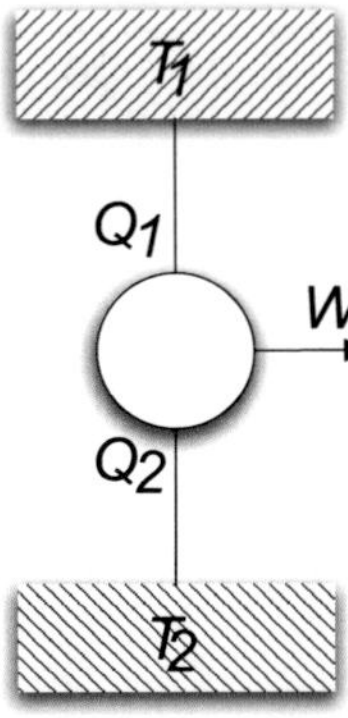

Figure 3.1. Bi-thermal engine operating between two sources at $T_1 > T_2$.

¿Could we know the sign of Q_1 and Q_2?

Applying the Second Law: Necessarily $Q_2 < 0$, which means it is liberated. Otherwise it would mean that work could be transformed into heat and transferred to the hot source without any other effect, which is against Clausius's statement, as shown in f gure 3.2.

Applying the First Law: $\Delta U = Q - W \Rightarrow W = Q = Q_1 + Q_2 > 0$. Hence, $Q_1 > -Q_2 > 0$

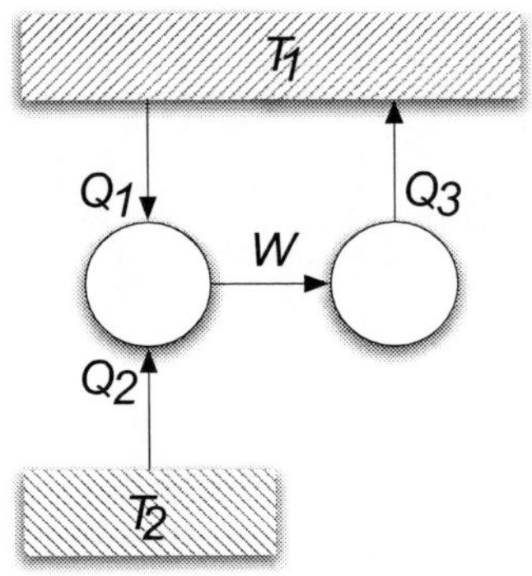

Figure 3.2. Setting which shows why it is necessary that the direct engine gives heat to the cold source.

The ratio between the produced work and the amount of heat absorbed from the hot source is the *eff ciency* of a direct engine

$$\eta = \frac{W}{Q_1} \tag{3.3}$$

3.5.2. Carnot's Theorem

All Carnot cycles that work between the same sources have equal eff ciency.

In order to understand this statement it is necessary to def ne a Carnot cycle. After this def nition, the theorem will be proven.

The Carnot Cycle

A Carnot cycle is a sequence of four reversible processes, two of which are at constant temperature and the other two adiabatic, as shown in f gure 3.3 for the case in which the only metric variable is volume. The direct cycle is followed clockwise and the reverse one

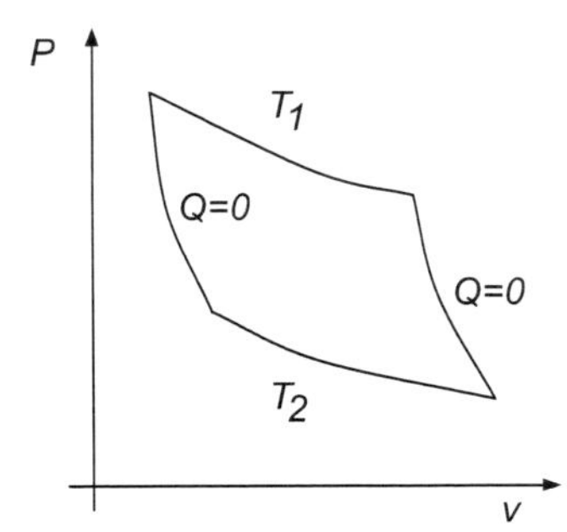

Figure 3.3. Example of a Carnot cycle.

counterclockwise. As it will be proven later, the Carnot cycle has the highest eff ciency of all the possible cycles operating between two given sources.

Proof of Carnot's Theorem

We intend to prove that any two direct bithermal engines that follow a Carnot cycle between the same sources have, necessarily, the same eff ciency.

PROOF

1. We shall imagine two engines R y R' that may be, in general, different, but that operate by following a Carnot cycle between these same sources (f gure 3.4). In spite of being different, they could always be made to produce the same amount of work, W.

> R and R' can be of different sizes, so that in principle, their work can be very different. But, as a process made by any number of cycles is itself a cycle, it will be always possible to consider a cycle made of the convenient number of cycles for each of the thermal machines, so that the work is the same.

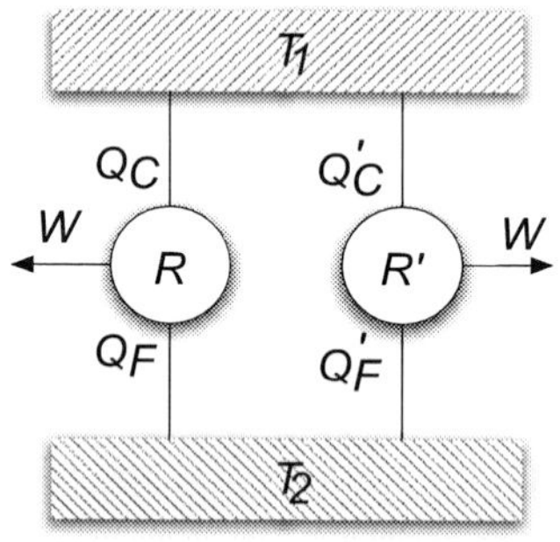

Figure 3.4. First step: Two bithermal Carnot engines, different in general, operating between the same sources and producing the same amount of work.

2. Being reversible engines, any of the two could be reversed so that it exchanged the same amounts of heat and work, but with opposite signs to those in f gure 3.4. We are going to analyze both possibilities.

3. First we reverse engine R' and we input work into it with R, as in f gure 3.5. Naturally, the amounts of heat exchanged by R' would have opposite signs to those in the original situation, while R operates the same.

4. Applying the First Law to the set $R+R'$:

$$\Delta U = \oint dU = 0 = Q - W = Q \Rightarrow$$

$$Q = Q_C - Q'_C + Q_F - Q'_F = 0 \Rightarrow Q_C - Q'_C = -Q_F + Q'_F$$

On the other hand, by the Clausius statement it is known that: $Q_C - Q'_C \geq 0$

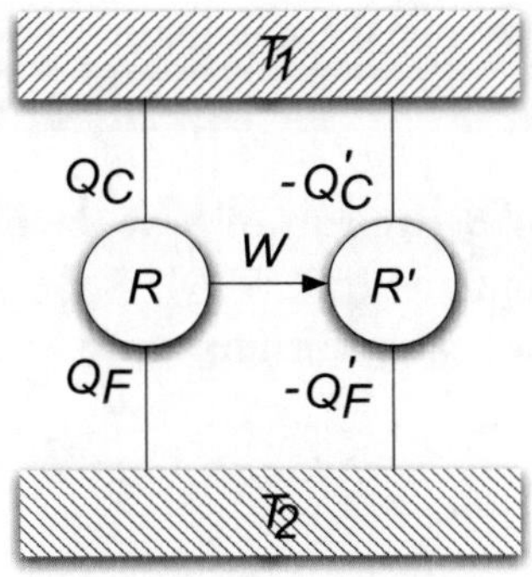

Figure 3.5. Second step: Engine R gives work to R', which works in reverse mode.

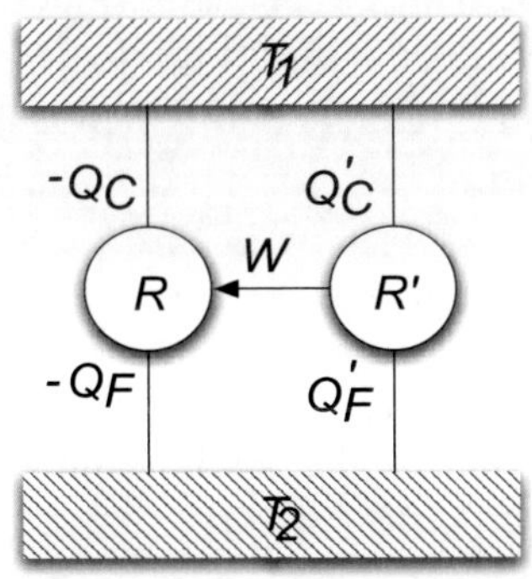

Figure 3.6. Third step: Engine R' gives work to R working in reverse mode.

5. Now, let us return to the original situation and reverse the operation of R, leaving R' in direct mode, as in f gure 3.6. If we analyze the new case:

$$\Delta U = 0 = Q - W = Q \Rightarrow$$
$$Q = -Q_C + Q'_C - Q_F + Q'_F = 0 \Rightarrow -Q_C + Q'_C = Q_F - Q'_F$$
$$\text{And by the Clausius statement: } -Q_C + Q'_C \geq 0$$

6. We have therefore arrived at the conclusion that the following must hold at the same time:

$$Q_C - Q'_C \geq 0$$
$$-Q_C + Q'_C \geq 0$$

This can only happen if $Q_C = Q'_C$ y $Q_F = Q'_F$.

7. Now, let us calculate the eff ciencies of both engines:

$$\eta_R = \frac{W}{Q_C} = \frac{Q_C + Q_F}{Q_C} = 1 + \frac{Q_F}{Q_C}$$
$$\eta_{R'} = \frac{W}{Q'_C} = \frac{Q'_C + Q_F}{Q'_C} = 1 + \frac{Q'_F}{Q'_C}$$
$$\text{Therefore: } \boxed{\eta_R = \eta_{R'}}$$

This equality holds for any two Carnot engines operating between sources 1 and 2.

As the previous reasoning is independent of how the machines might be built, their eff-ciency can only depend in a common trait which is the temperature of the sources. Therefore, the ratio of the amounts of heat exchanged must also be a function only of the temperatures of the sources:

$$\frac{Q_2}{Q_1} = \tau(t_2, t_1)$$

3.5.3. Existence of Thermodynamic Temperature

It can be observed that the validity of the previous deduction is not limited by the existence of any particular type of thermometer. This allows to def ne a new temperature, independent of the device to measure it (thermometer) and which is def ned in all conditions. It will be called *thermodynamic temperature.*

PROOF

1. According to Carnot's theorem, the ratio of the amounts of heat exchanged by a Carnot engine is a function of the temperatures of the sources exclusively: $\frac{Q_2}{Q_1} = \tau(t_2, t_1)$.

2. Let us consider the arrangement of engines of f gure 3.7, which includes three sources at empirical temperatures t_1, t_2 and t_3.

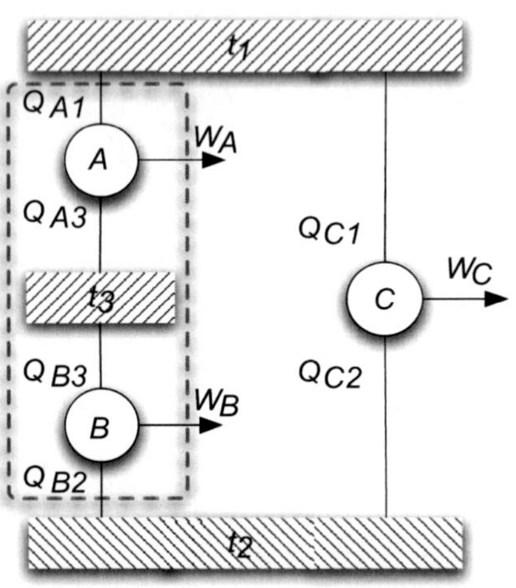

Figure 3.7. Arrangement of Carnot engines to deduce $\frac{T_F}{T_C} = -\frac{Q_F}{Q_C}$.

3. We shall assume

$$Q_{A1} = Q_{C1}$$
$$Q_{A3} = -Q_{B3}$$

4. The set of A+B+source 3 is also a Carnot engine operating between the sources at t_1 and t_2, as engine C, so:

$$\frac{Q_{A1}}{Q_{B2}} = \frac{Q_{C1}}{Q_{C2}} \Rightarrow Q_{B2} = Q_{C2}$$

5. Let us apply Carnot's theorem to machines A, B, and C:

$$\frac{Q_{A1}}{Q_{A3}} = \tau(t_1,t_3)$$
$$\frac{Q_{B3}}{Q_{B2}} = -\frac{Q_{A3}}{Q_{B2}} = \tau(t_3,t_2)$$
$$\frac{Q_{C1}}{Q_{C2}} = \frac{Q_{A1}}{Q_{B2}} = \tau(t_1,t_2)$$

6. Multiplying the f rst two and comparing with the last:

$$-\tau(t_1,t_3)\,\tau(t_3,t_2) = \frac{Q_{A1}}{Q_{B2}} = \tau(t_1,t_2)$$

7. This type of relation can only hold in the case that $\tau(t_1,t_2)$ has the mathematical form:

$$\tau(t_1,t_2) = -\frac{T(t_1)}{T(t_2)}$$

being T a function of the empirical temperature, which we shall call *thermodynamic temperature*.

Calculating the derivative by t_3

$$-\mathcal{T}(t_3,t_2)\frac{\partial \mathcal{T}(t_1,t_3)}{\partial t_3} - \mathcal{T}(t_1,t_3)\frac{\partial \mathcal{T}(t_3,t_2)}{\partial t_3} = 0 \Rightarrow$$
$$\frac{\partial \ln \mathcal{T}(t_1,t_3)}{\partial t_3} = -\frac{\partial \ln \mathcal{T}(t_3,t_2)}{\partial t_3}$$

The f rst member of this equality does not depend from t_2 and the second does not depend from t_1, so both must be a function only of t_3. By integrating, it results:

$$\ln \mathcal{T}(t_1,t_3) = \mathcal{F}(t_3) + \ln T(t_1) \Rightarrow \mathcal{T}(t_1,t_3) = T(t_1)\exp \mathcal{F}(t_3)$$
$$-\ln \mathcal{T}(t_3,t_2) = \mathcal{F}(t_3) + \ln T(t_2) \Rightarrow \frac{1}{\mathcal{T}(t_3,t_2)} = T(t_2)\exp \mathcal{F}(t_3)$$

and dividing, $-\mathcal{T}(t_1,t_2) = \frac{T(t_1)}{T(t_2)}$

Function $T(t)$ is not determined but its ratio is by the ratio of the amounts of exchanged heat of a Carnot engine, which could be obtained empirically.

In order to define the thermodynamic Kelvin scale from the previous, it is necessary to set the temperature value for a particular state. As with gas thermometers, this state is the triple point of water: $T_t = 273.16$ K.

3.6. Clausius' Equality and Inequality

In this section, we are going to introduce the concept of *entropy*, which will allow us to quantify the irreversibility of a process. For this, we shall first deduce Clausius' equality and inequality:

$$\oint \frac{dQ^R}{T} = 0 \qquad\qquad \oint \frac{dQ}{T} < 0$$

Clausius' equality — Clausius' inequality

Q represents the amount of heat exchanged by a system and T the temperature at which the exchange takes place. Along a cycle, the integral of $\frac{dQ}{T}$ will be equal to zero if the process carried out by the system is reversible, and less than zero otherwise.

PROOF

General description: In order to deduce Clausius' equality and inequality, we shall consider a closed, non-relativistic system, **A**, that exchanges heat at a certain temperature T and also exchanges work with its environment. We shall impose no further condition on it:

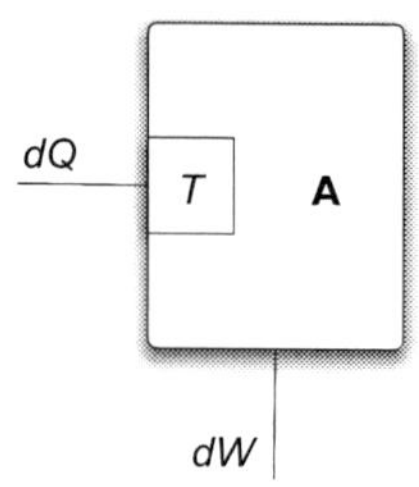

In order to eliminate any possible irreversibility which could occur in the heat and work exchanges, we shall insert Carnot engines and two sources which will interface system and environment instead of coupling them directly. The arrangement is shown in figure 3.8.

Now, we shall imagine a generic, cyclic process in system A with the arrangement shown in the figure, in order to calculate $\oint \frac{dQ}{T}$, expressing the result of the integral as a function of Q_2'. We shall then deduce its sign by applying Clausius' postulate.

1. Relation between Q_1, Q_1' and Q_2', obtained by applying the First Law to the boxed set, in the cycle carried out by A:

$$\Delta U = 0 = -Q_1 - Q_1' - Q_2' \Rightarrow Q_1 = -Q_1' - Q_2'$$

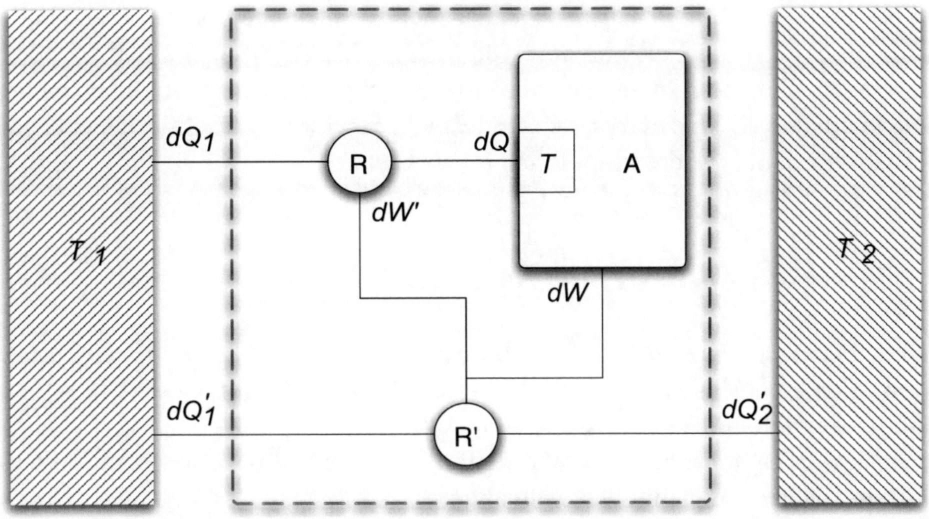

Figure 3.8. Arrangement for deducing Clausius' equality and inequality. dQ_1 y dQ_1' are the differential amounts of heat received by source 1, dQ_2' the one received by source 2. dW is the amount of work produced by A and dQ is the amount of heat it absorbs.

2. Applying Carnot's theorem to the two engines (and working out):

$$\frac{dQ_1}{T_1} + \frac{dQ}{T} = 0 \Rightarrow \oint \frac{dQ}{T} = -\frac{Q_1}{T_1}$$
$$\frac{dQ_1'}{T_1} + \frac{dQ_2'}{T_2} = 0 \Rightarrow \frac{Q_1'}{T_1} = -\frac{Q_2'}{T_2}$$

3. From the First Law and the previous:

$$\oint \frac{dQ}{T} = Q_2'\left(\frac{1}{T_1} - \frac{1}{T_2}\right)$$

4. It must be taken into account that:

$$T_1 > T_2 \Rightarrow \left(\frac{1}{T_1} - \frac{1}{T_2}\right) < 0$$

by Clausius' statement: $\quad Q_2' \geq 0$

5. So: $\boxed{\oint \frac{dQ}{T} \leq 0}$

6. If the process carried out by A was reversible, the heat f uxes involved in the process in one direction would be perfectly compensated for by the ones involved in the inverse. In that case, $\oint \frac{dQ^R}{T} = 0$. In the case that the process was not reversible, the compensation would not be perfect, hence $\oint \frac{dQ}{T} < 0$.

It must be observed that the Clausius equality requires the condition of A carrying out a *reversible* process. It must also be observed that, having substituted a real environment for an arrangement of Carnot engines, the only causes for irreversibility that may appear are those internal to A. This makes our conclusion a necessary but not suff cient condition for reversibility: it only addresses internal reversibility.

3.7. Entropy and Entropy Balance

We shall call *entropy f ux* to $dJ_s = \frac{dQ}{T}$. Clausius' equality, $\oint \frac{dQ^R}{T} = \oint dJ_s^R = 0$, proves that the entropy f ux coincides with the variation of a state function in case of closed systems performing reversible processes. This state function is *entropy*.

We are now going to deduce the balance of entropy for a closed, non-relativistic system, starting from Clausius' equality and inequality.

PROOF

Initial considerations: Let us consider a cycle containing two states 1 and 2, in which the process $1 \to 2$ is irreversible and $2 \to 1$ is reversible.

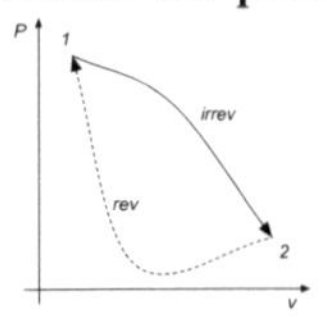

1. Because a part of the process is irreversible:

$$\oint \frac{dQ}{T} < 0$$

2. By decomposing the integral in stretches:

$$\oint \frac{dQ}{T} = \int_1^2 \frac{dQ}{T} + \underbrace{\int_2^1 \frac{dQ^R}{T}}_{-\Delta S_{1\to 2}} < 0 \Rightarrow \Delta S_{1\to 2} > \int_1^2 \frac{dQ}{T}$$

3. In order to make it an equality, the second member requires adding a strictly positive quantity, the internal entropy generation:

$$\Delta S = J_s + \sigma, \qquad \sigma > 0$$

4. By taking into account both reversible and irreversible processes, we f nally have:

$$\boxed{\Delta S = J_s + \sigma, \qquad \sigma \geq 0}$$

3.7.1. Some General Properties and Aspects Regarding S

S is a state function: S is a state function the same as U, H, P, T, V, etc. and therefore it verif es the properties of any state function. Its dimensions are those of energy divided by temperature, and therefore it is measured in J/K.

Origin of S: As in the case if U, only the increment ΔS for a closed system is def ned, and consequently, the origin of entropy is arbitrary. We shall see when we assess the Third Law of Thermodynamics that all substances share a common origin at $T = 0\,\text{K}$. However, it is the common practice to ignore this, to set the origin arbitrarily, and to make the necessary corrections when it becomes necessary to make different origins compatible.

S is additive: In a system formed by m subsystems, the entropy of the system is the addition of that of the subsystems, provided they are all calculated with compatible origins:

$$dS = \sum_{i=1}^{m} dS_i = \sum_{i=1}^{m} \left(\frac{dQ_i}{T_i} + d\sigma_i \right)$$

Apart from this, the entropy of the system as a whole, as in any other system, verif es:

$$dS = \sum_{i=1}^{m} \frac{dQ_i^{ex}}{T_i} + d\sigma$$

where ex indicates heat exchange with the exterior of the system for each of the i possible points at which it can take place.

Given that S is extensive, the *specif c entropy* is def ned as $s = \frac{S}{m}$.

Heat exchange: Taking into account $d\sigma = dS - dJ_s$, it can be verif ed that, in order for a process of heat exchange to be reversible, the two systems in thermal contact must have the same temperature.

3.7.2. Entropy Balance in Adiabatic Processes

The expression of the entropy balance for $Q = 0 \Rightarrow J_s = 0$ is:

$$\sigma = \Delta S^A \geq 0 \tag{3.4}$$

which indicates that the entropy of a system carrying out an adiabatic process *never decreases* (principle of entropy increase). It must be observed that the processes in which $\Delta S^A < 0$ are not possible and that, as a consequence, the states to which they get *are not reachable adiabatically*. This is an alternative formulation of the Second Law (see Caratheodory's statement in section 3.4.), however less intuitive and general than Clausius' and Kelvin's statements.

3.7.3. Entropy Balance in Compound Systems

In this section, we are going to analyze the forms in which entropy is generated in compound systems and we are going to distinguish *internal* and *external entropy generation.*

We are also going to work out entropy balances analyzing bi- and tri-thermal, non-reversible engines.

External Entropy Generation

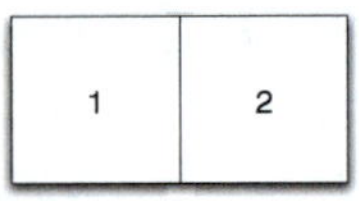

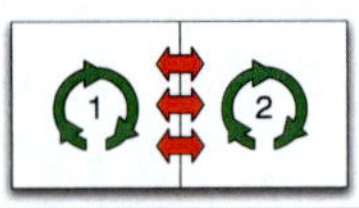

Let us consider a compound adiabatic system (see the f gure showing a simple two-system example). The subsystems, at T_1 and T_2, are not at thermal equilibrium so that they are exchanging heat between them.

We shall prove that the entropy generation in the compound system is the sum of: the internal entropy generation of each of the systems and an (external) entropy generation due to the irreversibility of the heat exchange between them.

PROOF

General description: We are going to combine the expressions for dS for the compound system (1) as a sum of each subsystem's dS on one side, and on the other, (2) as a system itself.

1. Additivity of entropy:

$$dS = \sum_i dS_i = \sum_i \left(\frac{dQ_i}{T_i} + d\sigma_i \right)$$

2. Entropy balance of the compound system:

$$dS = \sum_i \frac{dQ_i^{ex}}{T_i} + d\sigma$$

3. They must be equal:

$$\sum_i \left(\frac{dQ_i}{T_i} + d\sigma_i \right) = \sum_i \frac{dQ_i^{ex}}{T_i} + d\sigma \tag{3.5}$$

4. Let us analyze dQ_i. It is the sum of the heat exchanged by subsystem i with the environment and the heat it exchanges with the rest of subsystems:

$$dQ_i = dQ_i^{ex} + \sum_{j \neq i} dQ_{ij} \Rightarrow \frac{dQ_i}{T_i} = \frac{dQ_i^{ex}}{T_i} + \sum_{j \neq i} \frac{dQ_{ij}}{T_i}$$

The sum for all subsystems:

$$\sum_i \frac{dQ_i}{T_i} = \sum_i \frac{dQ_i^{ex}}{T_i} + \sum_i \sum_{j>i} dQ_{ij} \left(\frac{1}{T_i} - \frac{1}{T_j} \right)$$

5. By substituting in (3.5) and working out:

$$\sum_i \sum_{j>i} dQ_{ij}\left(\frac{1}{T_i}-\frac{1}{T_j}\right)+\sum_i d\sigma_i = d\sigma$$

6. We call *external generated entropy* in the heat exchange between two subsystems i and j to:

$$d\sigma^e_{ij} = dQ_{ij}\left(\frac{1}{T_i}-\frac{1}{T_j}\right)$$

According to Clausius' statement, if $T_i > T_j \Rightarrow dQ_{ij} < 0$ y si $T_i < T_j \Rightarrow dQ_{ij} > 0$,

$$\boxed{d\sigma^e_{ij} = dQ_{ij}\left(\frac{1}{T_i}-\frac{1}{T_j}\right) \geq 0} \tag{3.6}$$

where the equal sign is only valid for $T_i = T_j$ (and for the trivial case of no heat exchange). Hence,

$$\boxed{\sigma = \sum_i \sigma_i + \sum_i \sum_{j>i} \sigma^e_{ij}} \tag{3.7}$$

This reads that the internal entropy generation of a compound system can be expressed as the sum of the entropies generated internally in each of its subsystems and the external ones, due to heat exchanges taking place with a f nite temperature difference between subsystems.

Let us analyze a simple case: A compound adiabatic system formed by two subsystems at temperatures 1 and 2.

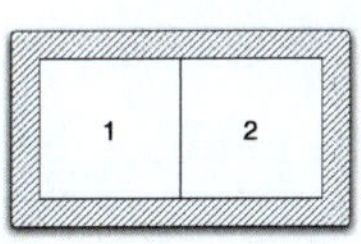

- Additivity: $dS = dS_1 + dS_2 = \frac{dQ_1}{T_1} + d\sigma_1 + \frac{dQ_2}{T_2} + d\sigma_2$
- dS of the compound system: $dS = \frac{\cancelto{0}{dQ_1^{ext}}}{T_1} + \frac{\cancelto{0}{dQ_2^{ext}}}{T_2} + d\sigma = d\sigma$
- Combining both ($dQ_2 = -dQ_1$):

$$0 \leq d\sigma = dQ_1\left(\frac{1}{T_1}-\frac{1}{T_2}\right)+d\sigma_1+d\sigma_2$$

Application to Bithermal Engines

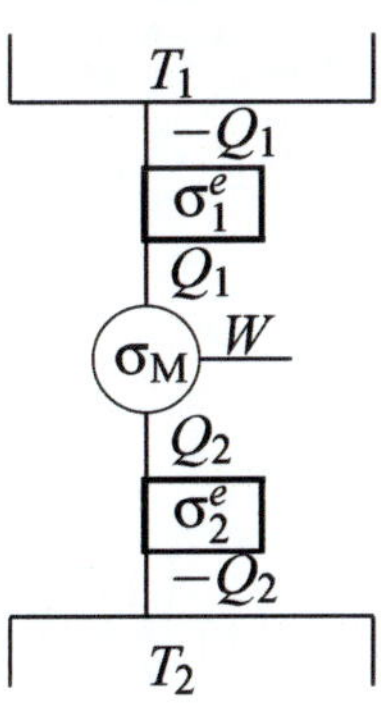

We are going to see how the generated entropy (irreversibility) affects the eff ciency of a bithermal engine operating *directly* for producing power and *reversely*.

Let us consider a bithermal engine as the one shown in f gure 3.1 p. 34.

In general, there will exist an internal irreversibility due to the engine itself and external irreversibilities in the heat exchanges with the sources 1 and 2. The processes in the sources are always internally reversible.

- Applying the First Law to the engine: $W = Q_1 + Q_2$
- Applying the Second Law to the set:

$$\text{Due to additivity:} \qquad \Delta S = \Delta S_M + \Delta S_{F1} + \Delta S_{F2} = 0 + \frac{-Q_1}{T_1} + \frac{-Q_2}{T_2}$$

$$\text{As a compound system:} \qquad \Delta S = \sigma = \sigma_M + \sigma_1^e + \sigma_2^e \geq 0$$

- Thus, Q_1 and Q_2 can be expressed as a function one of the other:

$$Q_1 = -Q_2 \frac{T_1}{T_2} - T_1 \sigma$$

$$Q_2 = -Q_1 \frac{T_2}{T_1} - T_2 \sigma$$

- We substitute in the First Law in order to work out W:

$$W = Q_1 + Q_2 = Q_1 \left(1 - \frac{T_2}{T_1}\right) - T_2 \sigma = Q_2 \left(1 - \frac{T_1}{T_2}\right) - T_1 \sigma$$

- For the direct engine, the eff ciency results: $\eta = \frac{W}{Q_1}$:

$$\eta = \underbrace{\left(1 - \frac{T_2}{T_1}\right)}_{\eta_R} - \underbrace{\frac{T_2}{Q_1}\sigma}_{\geq 0} \leq \eta_R < 1$$

It can be observed that the eff ciency of a bithermal engine is always less than that of a reversible one which would operate between the same sources: The Carnot engine has the *greatest* eff ciency.

- With reverse engines, two cases must be considered:

1. When the desired effect is the amount of heat extracted from the cold source, the engine is called *refrigerator* and its eff ciency is:

$$\varepsilon = \frac{Q_2}{-W} = \frac{1}{-1 + \frac{T_1}{T_2} + \frac{T_1 \sigma}{Q_2}} \leq \frac{1}{-1 + \frac{T_1}{T_2}} = \frac{T_2}{T_1 - T_2} = \varepsilon^R$$

again, the eff ciency of the refrigerating Carnot cycle is the greatest possible. It must be observed that the eff ciency of a refrigerating cycle can be less or greater than 1.

2. When the desired effect is the amount of heat received by the hot source, the engine is called *heat pump*. Its eff ciency is

$$\left.\begin{array}{l} \varepsilon = \dfrac{-Q_1}{-W} = \dfrac{Q_1}{Q_1 + Q_2} > 1 \\[2ex] \varepsilon = \dfrac{1}{1 - \frac{T_2}{T_1} - T_2 \frac{\sigma}{Q_1}} \leq \dfrac{1}{1 - \frac{T_2}{T_1}} = \dfrac{T_1}{T_1 - T_2} = \varepsilon^R \end{array}\right\} \Rightarrow 1 < \varepsilon \leq \varepsilon^R$$

The eff ciency of a bithermal heat pump is always greater than 1, and less than that of a reversible one operating between the same sources.

Tri-thermal Engine

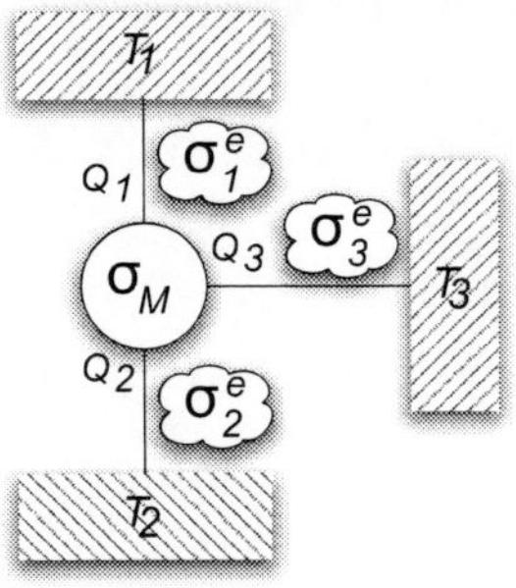

We shall only consider the case of an *absorption refrigerator* (see the f gure). Work is negligible and source 3 is the hottest so that : $T_3 > T_1 > T_2$. Source 2 is the target desired to be refrigerated, and 1 is the environment.

- Applying the First Law: $Q_1 + Q_2 + Q_3 = 0 \Rightarrow Q_1 = -(Q_2 + Q_3)$
- The Second Law to the set:

$$\frac{Q_1}{T_1} + \frac{Q_2}{T_2} + \frac{Q_3}{T_3} = -\sigma$$

- Now, substituting the First Law into the Second:

$$Q_2\left(\frac{1}{T_2} - \frac{1}{T_1}\right) + Q_3\left(\frac{1}{T_3} - \frac{1}{T_1}\right) = -\sigma$$

- Working out: $Q_2 = Q_3 \frac{T_2}{T_3}\frac{T_3 - T_1}{T_1 - T_2} - \sigma\frac{T_1 T_2}{T_1 - T_2}$
- Eff ciency, in this case is def ned as:

$$\varepsilon = \frac{Q_2}{Q_3} = \frac{T_2}{T_3}\frac{(T_3 - T_1)}{(T_1 - T_2)} - \frac{\sigma}{Q_3}\frac{T_1 T_2}{T_1 - T_2} \leq \frac{T_2(T_3 - T_1)}{T_3(T_1 - T_2)} = \varepsilon^R$$

Again, eff ciency is less than in the reversible case.

> It must be observed that the eff ciency of the reversible trithermal engine tends to that of the bithermal reversible one as $T_3 \to \infty$. The meaning of this will appear more clearly when we study exergy: in a refrigerating engine, it is essential to input exergy, which can be done by giving work to the engine (the normal idea) or by giving it exergy through the exergy content of the heat Q_3.

3.7.4. Increment of Entropy in Non-static Processes

In non-static processes, intermediate states are not def ned. We are going to deduce that it holds, for two, initial and f nal states:

$$\Delta S_{12} = S_2 - S_1 > \sum_i \frac{Q_i}{T_i}$$

Where i is each of the sources with which the system may exchange heat.

- The variation of entropy of a source is: $\Delta S_F = -\frac{Q_F}{T_F}$
  being $-Q_F$ the amount of heat absorbed by the source, and T_F its temperature.
- Entropy balance applied to the universe of the system:

$$\Delta S_u = S_2 - S_1 + \sum_i \frac{-Q_i}{T_i} > 0 \Rightarrow \boxed{S_2 - S_1 > \sum_i \frac{Q_i}{T_i}}$$

3.8. Gibbs' Equation and Jacobian Relation

In this section, we are going to introduce *Gibbs' equation* and the Jacobian relation. We shall restrict ourselves to the case of systems whose only metric variable is the volume. Systems with more metric variables can be treated in a similar way. We have

$$T\,dS = dU + P\,dV \qquad \frac{\oint T\,dS}{\oint P\,dV} = 1$$

Gibbs' equation Jacobian relation

The f rst allows to relate entropy and internal energy, as well as the rest of state variables, and the second establishes a relation between the caloric coeff cients and the thermal ones.

Both will be necessary for calculating partial derivatives.

3.8.1. Gibbs' Equation

PROOF

1. On one side, there is the caloric equation of a system, $U = U(T,V)$
2. On the other, entropy, as a state function, can be expressed as $S = S(T,V)$
3. We could eliminate T, having:
$$U = U(S,V)$$
4. This expression represents the surface of equilibrium states. In order to obtain it in differential form, *any* elementary process on it may be considered. For simplicity, we consider a reversible one, so that it holds: $dS = \dfrac{dQ^R}{T}$
5. In a system whose only metric variable were volume, $dW^R = P\,dV$; and by the First Law: $dQ^R = dU + dW^R = dU + P\,dV$
6. Substituting and rearranging:

$$\boxed{dU = T\,dS - P\,dV} \quad \text{Gibbs' First Equation} \tag{3.8}$$

Taking into account that $dH = dU + P\,dV + V\,dP$ it follows:

$$\boxed{dH = T\,dS + V\,dP} \qquad \text{Gibbs' Second Equation} \tag{3.9}$$

3.8.2. Jacobian Relation

PROOF

1. We part from Gibbs' equation written as: $T\,dS = dU + P\,dV$
2. We integrate along a cycle: $\oint T\,dS = \cancelto{0}{\oint dU} + \oint P\,dV$
3. Rearranging the expression:

$$\oint T\,dS = \oint P\,dV \Leftrightarrow \frac{\oint T\,dS}{\oint P\,dV} = 1 \Leftrightarrow \boxed{\frac{\partial(TS)}{\partial(PV)} = 1} \tag{3.10}$$

it is the one we shall use for calculating derivatives.

3.8.3. Caloric Coeff cients as a Function of the Thermal Equation

The previous allows re-expressing coeff cients l_θ and λ_θ as a function of α, β y κ:

$$\begin{aligned} l_\theta &= \left(\frac{\partial u}{\partial v}\right)_T = T\left(\frac{\partial s}{\partial v}\right)_T - P = T\frac{\partial(sT)}{\partial(vT)} - P = T\frac{\partial(vP)}{\partial(vT)} - P = \\ &= T\left(\frac{\partial P}{\partial T}\right)_v - P = TP\beta - P = \boxed{P(T\beta - 1)} \end{aligned} \tag{3.11}$$

$$\lambda_\theta = \left(\frac{\partial h}{\partial P}\right)_T = T\left(\frac{\partial s}{\partial P}\right)_T + v = T\frac{-\partial(Pv)}{\partial(PT)} + v = \boxed{v(1 - T\alpha)} \tag{3.12}$$

It can be observed that the caloric coeff cients l_θ and λ_θ can always be expressed in terms of the thermal ones. In other words, knowing the thermal equation, l_θ and λ_θ are also known. From these expressions, it can be easily proven that both caloric coeff cients are zero for an ideal gas.

3.8.4. Thermodynamic Temperature and Ideal Gas Temperature

The previous relations allow determining the coeff cients as a function of any concept of temperature, and therefore, they also allow relating different concepts of temperature between themselves. This allows to check that, within the range in which the ideal gas temperature is def ned, it coincides with the thermodynamic temperature. Thus, ideal gas thermometers measure thermodynamic temperature directly (to prove this, solve $1/T$ from the f rst of these expressions, integrate dT/T between some arbitrary limits and take the limit at null pressure in the second member).

They also allow to obtain the thermodynamic temperature from experimental measurements based in any other empirical temperature, eliminating the need of going through the ideal gas temperature (this is best accomplished solving $1/T$ from the expression for the Joule-Thomson coeff cient, and integrating between known values).

We don't give any detailed proof of these relations, because they are not really interesting to engineers, but the results are worth mentioning in the classroom.

3.9. Variation of Entropy of an Ideal Gas

We are going to apply Gibbs' equations to a system formed by an ideal gas, in order to calculate Δs as a function of already known parameters.

PROOF

1. We part from Gibbs' equations: $T\,ds = du + P\,dv = dh - v\,dP$
2. We apply the already known expressions du y dh for an ideal gas:

$$du = c_v^* dT + l_\theta^* dv = c_v^* dT$$
$$dh = c_p^* dT + \lambda_\theta^* dP = c_p^* dT$$

3. We substitute in Gibbs' equations:

$$T\,ds = du + P\,dv = c_v^* dT + P\,dv^*$$
$$T\,ds = dh - v\,dP = c_p^* dT - v^*\,dP$$

4. Rearranging:

$$\boxed{ds = \frac{c_v^*}{T} dT + \frac{R}{v^*} dv^*} \qquad \boxed{ds = \frac{c_p^*}{T} dT - \frac{R}{P} dP}$$

3.10. Exergy

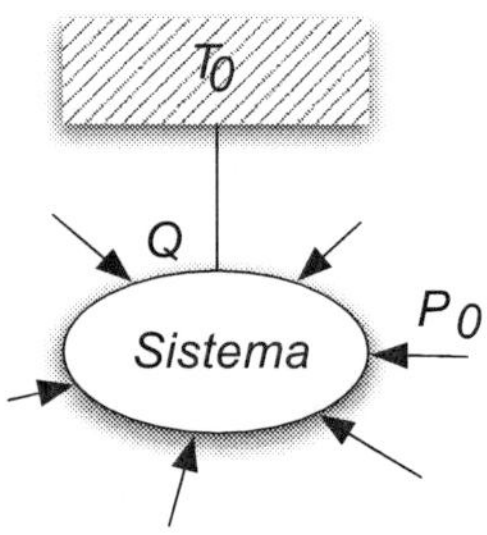

DEFINITION: Exergy is a state function that measures the maximum useful work that could be obtained from the closed system at rest, interacting only with the environment.

$$\boxed{B = U - U_0 - T_0(S - S_0) + P_0(V - V_0)} \qquad (3.13)$$

In moving systems, *total exergy* plays the same role: $B' = B + E_c + E_p$

PROOF

1. The useful work results from substracting the work of variation of volume against the environment from the total work:

$$W_{\text{useful}} = W - P_0 \Delta V \tag{3.14}$$

2. Let us relate it to the system's state variables:

$$\text{First Law:} \qquad W = Q - \Delta U'$$

$$\text{Second Law to the universe:} \qquad \Delta S_u = \Delta S + \frac{-Q}{T_0} \geq 0$$

3. Working out: $Q = T_0 \Delta S - T_0 \Delta S_u$

4. Substituting

$$W = Q - \Delta U' = T_0 \Delta S - T_0 \Delta S_u - \Delta U' \Rightarrow$$
$$W_{\text{useful}} = T_0 \Delta S - T_0 \Delta S_u - \Delta U' - P_0 \Delta V$$

5. In order for it to be the maximum useful work, the process must be reversible $\Delta S_u = 0$ thus:

$$W_{\text{useful}}^{\max} = T_0 \Delta S - \Delta U' - P_0 \Delta V = -(\Delta U' + P_0 \Delta V - T_0 \Delta S)$$

6. That is, that the $W_{\text{useful}}^{\max}$ that can be obtained from a closed system interacting only with the environment is equal to a linear combination of state functions, and, therefore, to the variation of another state function. This is the function we call *total exergy*:

$$W_{\text{useful}}^{\max} = -\Delta B' = -(\Delta U' + P_0 \Delta V - T_0 \Delta S)$$

7. If we call *dead state* to that in which no more work can be obtained from the system by interacting only with the environment, the maximum capacity to produce useful work is, evidently, the variation of B' with respect to this state. The temperature of the system at the dead state will necessarily be T_0, for otherwise, a thermal engine could work between system and environment obtaining work. Analogously, the pressure of the system must be P_0, for otherwise work could be obtained by placing a piston between the system and environment. For systems whose only metric variable were volume, this would def ne the dead state completely. By denoting the properties of the system at the dead state with the subindex '0', we have:

$$B' = (U' - T_0 S + P_0 V) - (U_0' - T_0 S_0 + P_0 V_0) \tag{3.15}$$

8. Exergy results from $B = B' - (E_c + E_p)$, so

$$\boxed{B = U - U_0 - T_0(S - S_0) + P_0(V - V_0)} \tag{3.16}$$

3.10.1. Properties of Exergy

- Both total exergy and exergy are state functions. They have dimensions of energy.
- The origin of exergy is environment conditions: (T_0, P_0).
- Exergy is different if the environment is different; it is not a mere change of its origin; the shape of the function changes (derivatives, for example.).
- The origin of total exergy is (T_0, P_0) and null kinetic and potential energies.
- The minimum possible value of exergy is 0, which can only happen at the dead state. At any other state, $B > 0$.

Unlike energy, exergy can be destroyed.

3.10.2. Exergy Balance

The previous expression gives exergy as a state function. By applying the First and Second Laws, it is straightforward to obtain an expression for the exergy balance as an alternative formulation for the Second Law stated in energetic terms.

$$\left.\begin{array}{r}\Delta U' = Q - W \\ \Delta S = J_s + \sigma\end{array}\right\} \Rightarrow \Delta B' = Q - W - T_0(J_s + \sigma) + P_0 \Delta V$$

which can be rewritten as:

$$\Delta B' = (Q - T_0 J_s) - (W - P_0 \Delta V) - I \qquad (3.17)$$

The physical meaning of each of the terms in the second member is:

$Q - T_0 J_s = \int dQ \left(1 - \frac{T_0}{T}\right)$ this is, for mono-thermal systems, the exergetic content of heat. The exergetic content of a certain amount of heat $\bar{d}Q$ has its same sign if $T > T_0$, and opposite if $T < T_0$. It must also be observed that the exergetic content of a certain amount of heat tends to the amount itself as $T \to \infty$, which explains the result that was obtained when studying the trithermal engine.

$W - P_0 \Delta V = W_{\mathbf{useful}}$ expresses the exergetic content of work.

$I = T_0 \sigma$ is the so-called *Gouy-Stodola equation*. I is the *exergetic destruction* caused by generated entropy σ, and it equals the loss of capacity of the system to produce work due to internal irreversibilities.

The *total exergy destruction* or *universe exergy destruction* is def ned as:

$$\boxed{I_t = T_0 \Delta S_u} \qquad (3.18)$$

which is the exergetic destruction that appears in the exergy balance, when applied to the universe of the system, which includes all causes for irreversibility, both internal and external.

Exercise 1: A cylinder with a moving piston contains 5 mol of an ideal gas of $c_v = 20$ J/mol K, at $P_1 = 20$ bar, $T_1 = 500$ K. The gas performs an isothermal quasiestatic expansion with friction, with a f nal volume of $V_2 = 4V_1$ absorbing 25 kJ of heat from a source at 1 000 K. Find:

1. Work performed in the process.
2. Frictional work.
3. Entropy change in the system, entropy f ux and internal entropy generation.
4. Change in the entropy of the universe.
5. External entropy generation.
6. Maximum work that could have been obtained from the system in the given change of state, interacting only with the ambient.
7. Exergy change.
8. Exergy destruction in the system and total exergy destruction.

Ambient state: 300 K and 1 bar.

Solution: Applying the First Law:

$$W = -\Delta U + Q = Q = 25\,\text{kJ}$$

$$W_d = W - \int_1^2 P\,dV = -nRT \ln\frac{V_2}{V_1} + W =$$

$$= -5\,\text{mol}\; 8.314\,\frac{\text{J}}{\text{mol K}}\; 500\,\text{K} \ln 4 + 25\,\text{kJ} = -3.81\,\text{kJ}$$

$$S_2 - S_1 = nR \ln\frac{V_2}{V_1} = 5\,\text{mol}\; 8.314\,\frac{\text{J}}{\text{mol K}} \ln 4 = 57.63\,\text{J/K}$$

$$J_s = \frac{Q}{T} = \frac{25\,000\,\text{J}}{500\,\text{K}} = 50\,\text{J/K} \Rightarrow \sigma = 57.63\,\text{J/K} - 50\,\text{J/K} = 7.63\,\text{J/K}$$

$$\Delta S_u = \Delta S + \Delta S_F = 57.63\,\text{J/K} + \frac{-25\,000\,\text{J}}{1\,000\,\text{K}} = 32.63\,\text{J/K}$$

$$\sigma^e = \Delta S_u - \sigma = 32.63\,\text{J/K} - 7.63\,\text{J/K} = 25\,\text{J/K}$$

$$W_{\text{max}} = -\Delta U + T_0 \Delta S = 300\,\text{K}\; 57.63\,\text{J/K} = 17.29\,\text{kJ}$$

$$\Delta B = \Delta U - T_0 \Delta S + P_0 \Delta V = -300\,\text{K}\; 57.63\,\text{J/K} + 1\,\text{bar}\,(4V_1 - V_1) =$$

$$= -17.29\,\text{kJ} + 1\,\text{bar}\frac{3 \times 5\,\text{mol}\; 8.314\,\frac{\text{J}}{\text{mol K}}\; 500\,\text{K}}{20\,\text{bar}} = -14.17\,\text{kJ}$$

$$I = T_0 \sigma = 300\,\text{K}\; 7.63\,\text{J/K} = 2.29\,\text{kJ}$$

$$I_t = T_0(\sigma + \sigma^e) = T_0 \Delta S_u = 9.79\,\text{kJ}$$

Chapter 4

Thermodynamic Potentials

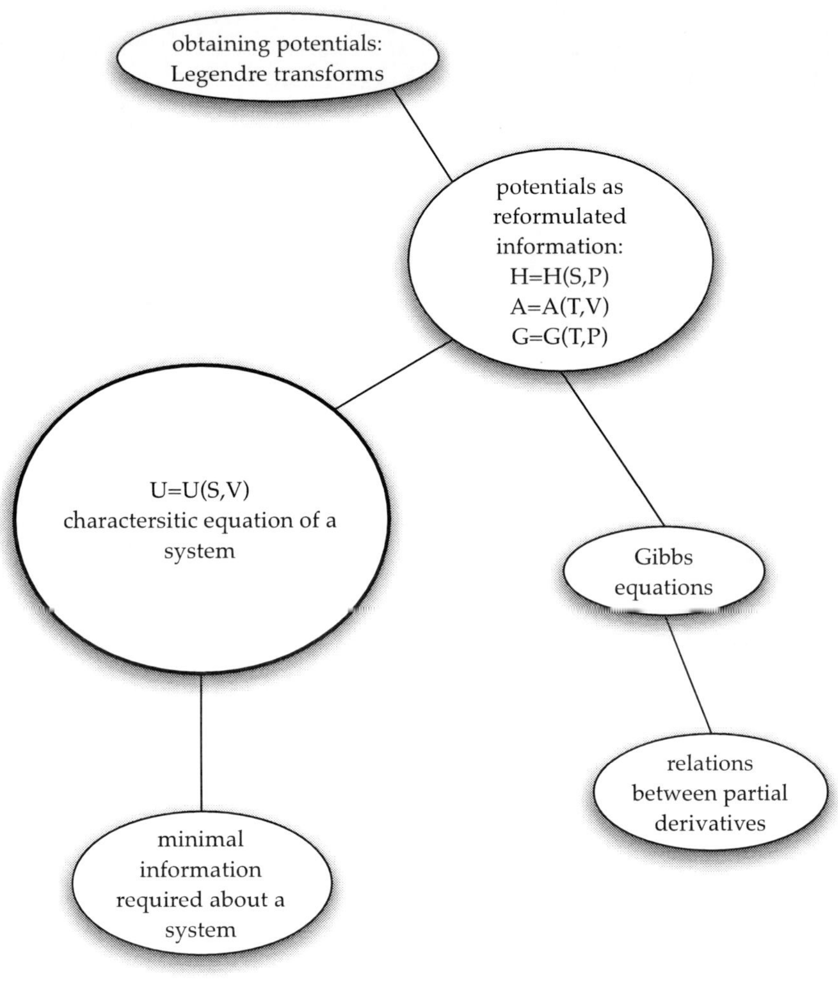

If we knew the correlation between internal energy, entropy, and volume, U, S, and V, we would know all there was to know –thermodynamically speaking– about a system. However, measuring S is impossible and measuring V is diff cult sometimes and sometimes impossible, so it appears that knowing how U, S, and V relate will not come straightforward.

However, information can frequently be expressed in a different, more immediate way, and this is exactly what happens with $U = U(S,V)$. There exists a mathematic transform which will turn the independent variables from the unmeasurable (S,V) to the more accessible (T,V) or the ideal (T,P), when U changes to be the Helmholtz function, A, or Gibbs' free energy, G. Of course, we had already mentioned $H = H(S,P)$.

4.1. Thermodynamic Potentials

DEFINITION: The potentials H, A, and G are def ned as

$$H = U + PV \quad ; \quad A = U - TS \quad ; \quad G = U + PV - TS$$

4.1.1. $U(S,V)$ Is a Characteristic Equation

We say that $U(S,V)$ is a *characteristic equation* of the system because it contains all the system's thermodynamic information. That is equivalent to say that it is possible to obtain from it the thermal and also the caloric and entropic equations of state, as we shall see:

Thermal equation:

$$U(S,V) \Rightarrow \left\{ \begin{array}{c} \left(\frac{\partial U}{\partial S}\right)_V = T(S,V) \\ \left(\frac{\partial U}{\partial V}\right)_S = -P(S,V) \end{array} \right\} \Rightarrow \text{eliminating } S\text{: } P = P(T,V)$$

Caloric equation:

$$\left. \begin{array}{r} \left(\frac{\partial U}{\partial S}\right)_V = T(S,V) \\ U = U(S,V) \end{array} \right\} \Rightarrow \text{eliminating } S\text{: } U = U(T,V)$$

Entropic equation:

$$\left(\frac{\partial U}{\partial S}\right)_V = T(S,V) \Rightarrow \text{isolating } S\text{: } S = S(T,V)$$

4.1.2. Determination of the Thermodynamical Potentials

Although the knowledge of $U(S,V)$ makes possible to obtain all the system's information, this equation has signif cant disadvantages:

- U cannot be measured directly.
- S cannot either.
- V can be measured, but there are practical problems that make it diff cult to be measured with precision.

Then the problem emerges of how to represent the *same information* contained in $U(S,V)$ using different variables. This problem is solved by the *Legendre transformation*.

Legendre transformation. Let us consider a function $Z = Z(X_1, X_2, \ldots X_m)$. To calculate the k-th transformation, select k variables X_i, $i = 1, \ldots, k$. The transformed function, Z^k, will depend on the k *conjugate* variables y_i of the selected ones and not on the selected original variables: X_j, $j = k+1, \ldots, m$. It is def ned as

$$Z^k(y_1, y_2, \ldots y_k, X_{k+1}, \ldots, X_m) = Z - \sum_{i=1}^{k} y_i X_i,$$

and the conjugate variables def nition is $y_i = \left(\frac{\partial Z}{\partial X_i}\right)_{X_{j \neq i}}$

Substituting V:

$$\left(\frac{\partial U}{\partial V}\right)_S = -P \Rightarrow Z^1(S,P) = \boxed{H(S,P) = U + PV} \quad \text{enthalpy} \tag{4.1}$$

Substituting S:

$$\left(\frac{\partial U}{\partial S}\right)_V = T \Rightarrow Z^1(T,V) = \boxed{A(T,V) = U - TS} \quad \text{Helmhotz function} \tag{4.2}$$

Substituting both S and V:

$$\left.\begin{array}{l} \left(\frac{\partial U}{\partial S}\right)_V = T \\ \left(\frac{\partial U}{\partial V}\right)_S = -P \end{array}\right\} \Rightarrow Z^2(T,P) = \boxed{G(T,P) = U + PV - TS} \quad \text{Gibbs function} \tag{4.3}$$

Legendre transformation keeps all the physical information and leads to a characteristic equation involving the potential G, which only depends on the intensive and easily measurable variables T and P.

4.1.3. The Gibbs Equations

From the potentials' def nitions and taking into account the f rst Gibbs equation, obtained previously, the four Gibbs equations for closed systems with a single metric variable (the volume) are straightforward:

$$dU = \quad T\,dS - P\,dV \tag{4.4}$$

$$dH = \quad T\,dS + V\,dP \tag{4.5}$$

$$dA = \quad -S\,dT - P\,dV \tag{4.6}$$

$$dG = \quad -S\,dT + V\,dP \tag{4.7}$$

4.1.4. Minimum Information

As we can observe in the derivation tree of the potential, that we shall see in the following sections, the minimum information is at the same time easily measurable and equivalent to the characteristic equation, is:

- The thermal equation of state.
- The $c_p = c_p(T,P)$ along an isobar or the $c_v = c_v(T,v)$ along an isochore; e.g., *the ideal gas limit* isobar and isochore, $c_p(T)|_{P=0} = c_p^*(T)$, or $c_v(T)|_{v=\infty} = c_v^*(T)$.

4.2. Partial Derivatives and Minimum Information

As stated before, $U = U(S,V)$, $H = H(S,P)$, $A = A(T,V)$, and $G = G(T,P)$ are characteristic equations. From any of them, the system's equations of state can be obtained in an analogue form to what was seen for the f rst one. However, this description of the system presents the diff culty of measuring the potentials conveniently, since this is not easy at all.

4.2.1. Derivation Tree of Potentials

The following derivation tree of the potential $g(T,P)$ could be easily obtained for any other potentials, too. It illustrates how all the properties of the system can be obtained, integrating, deriving and applying the cross-derivatives equality properly.

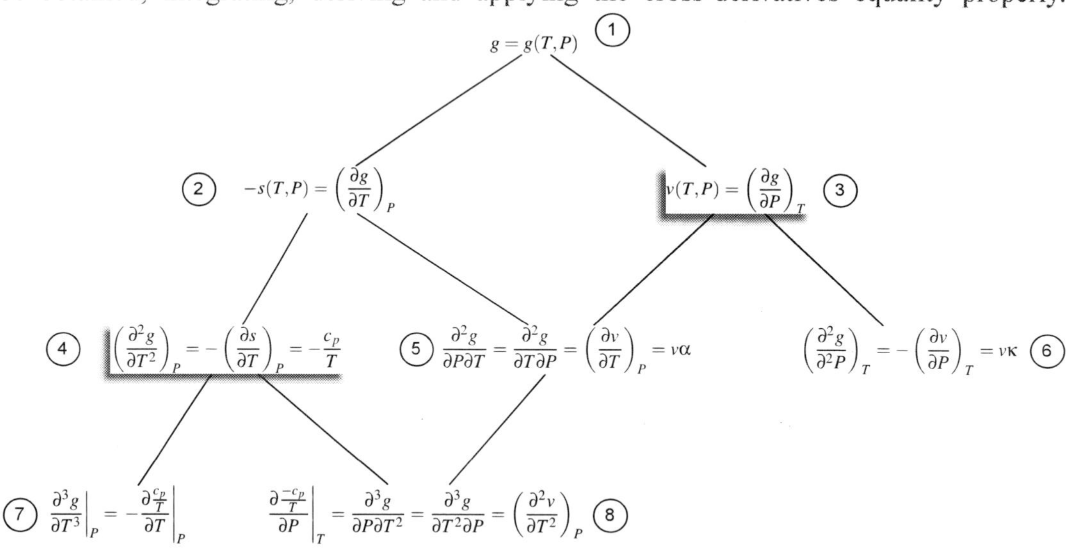

From the thermal equation $v\alpha$ is immediately got, what, together with c_p, allows to obtain $s(T,P)$ by integration (the constant of integration is f xed by setting the arbitrary reference state for entropy).

Also, it can be observed that $c_p(T,P)$ can be obtained from the knowledge of the function $c_p^*(T) = c_p(T,0)$ and the thermal equation (here there is not constant of integration because the integral is def ned), since $(\partial c_p/\partial P)_T$ could be calculated from it. Finally, we could integrate the entropic equation together with the thermal equation in order to obtain the potential $g = g(T,P)$; here the constant of integration f x the arbitrary state of reference for g.

Inspecting the derivation tree and taking into account the cross-derivatives equality of state functions, multiple relations between the partial derivatives are obtained. As an example, the so-called Maxwell relations arise from (4.6) and (4.7):

$$\left(\frac{\partial s}{\partial P}\right)_T = -\left(\frac{\partial v}{\partial T}\right)_P \quad ; \quad \left(\frac{\partial s}{\partial v}\right)_T = \left(\frac{\partial P}{\partial T}\right)_v \tag{4.8}$$

4.2.2. Relations between Partial Derivatives

The number of partial derivatives that can be written (which may be interesting to know for any specif c application) is obviously huge; however, all of them can be obtained without any more information than the minimum cited before (and the particular states of reference to be taken for those functions that have an arbitrary origin). This shows that *any* partial derivative can be related to the derivativeds of the thermal equation and c_v (alternatively c_p). Remember that the last one is obtained simply by knowing c_v^* (or c_p^*) and the thermal equation of state, through the immediate relations

$$c_v = c_v^* + \int_\infty^v \left(\frac{\partial c_v}{\partial v}\right)_T dv \quad ; \quad c_p = c_p^* + \int_0^P \left(\frac{\partial c_p}{\partial P}\right)_T dP$$

where the two derivatives displayed are obtained by applying the crossed-derivatives equality to any equation involving the corresponding specif c heat and whose independent variables are T and v (or P). The resulting expression are called Clausius' relations.

As an example, for c_v it would be worth to use the characteristic equation $a = a(T,v)$, the caloric equation $u = u(T,v)$ or the entropic equation $s = s(T,v)$. We will deduce the corresponding Clausius' relation from the last one:

$$ds = \left(\frac{\partial s}{\partial T}\right)_v dT + \left(\frac{\partial s}{\partial v}\right)_T dv$$

$$\left(\frac{\partial s}{\partial T}\right)_v = \frac{c_v}{T} \quad \text{(as we will see later)} \quad ; \quad \left(\frac{\partial s}{\partial v}\right)_T = \left(\frac{\partial P}{\partial T}\right)_v \quad \text{(Maxwell relations)}$$

$$\text{Crossed-derivatives equality:} \quad \frac{1}{T}\left(\frac{\partial c_v}{\partial v}\right)_T = \left(\frac{\partial^2 P}{\partial T^2}\right)_v \Rightarrow c_v = c_v^* + \int_\infty^v T\left(\frac{\partial^2 P}{\partial T^2}\right)_v dv$$

In other words, the problem is how to express any partial derivative exclusively in terms of the state variables α, β, κ, c_v, and c_p. Obviously, one of the f rst three and one of the two last ones can be avoided, due to the previously seen relation $\alpha/\kappa = \beta P$ (1.9), and $c_p = c_v + \alpha^2 vT/\kappa$ (4.9), that will be obtained as an exercise in page 60.

The procedure is summarized as follows:

1. Elimination of the potentials, one by one, expressing them as a function of the independent variables in their characteristic equations and applying the chain rule. By doing so, the partial derivatives of any potential will always be known by referring to the corresponding Gibbs equation.

2. Remember that if you have to eliminate the "outside variable" (i.e., the one that remains constant for the derivation), use the mathematical property

$$\left(\frac{\partial x}{\partial y}\right)_z\left(\frac{\partial y}{\partial z}\right)_x\left(\frac{\partial z}{\partial x}\right)_y=-1\Rightarrow\left(\frac{\partial y}{\partial z}\right)_x=-\frac{\left(\frac{\partial x}{\partial z}\right)_y}{\left(\frac{\partial x}{\partial y}\right)_z}$$

3. Once all potentials have been eliminated, it is entropy's turn. Any possible derivative involving s may be:

   - As already mentioned earlier, when s is constant, the property of partial derivatives product must be used.
   - Derivatives with respect to T: they come directly from the f rst two Gibbs equations (4.4) and (4.5).

$$ds=\frac{1}{T}du+\frac{P}{T}dv\Rightarrow\left(\frac{\partial s}{\partial T}\right)_v=\frac{c_v}{T}$$
$$ds=\frac{1}{T}dh-\frac{v}{T}dP\Rightarrow\left(\frac{\partial s}{\partial T}\right)_P=\frac{c_p}{T}$$

   - Derivatives with T constant: they are obtained from Maxwell relations (4.8).
   - Derivative with respect to P or v, with the other one constant: they are reduced to the previous cases "introducing T in the numerator and denominator" (this is just a particular use of the chain rule):

$$\left(\frac{\partial s}{\partial v}\right)_P=\left(\frac{\partial s}{\partial T}\right)_P\left(\frac{\partial T}{\partial v}\right)_P \quad ; \quad \left(\frac{\partial s}{\partial P}\right)_v=\left(\frac{\partial s}{\partial T}\right)_v\left(\frac{\partial T}{\partial P}\right)_v$$

Exercise 1: Find $c_p - c_v$ for any f uid as a function of α, κ and the variables of the thermal equation.

Solution From (2.6), that can be written as

$$c_p-c_v=\alpha v(l_\theta+P)$$

and (3.11):

$$l_\theta=P(T\beta-1)$$

we get

$$c_p-c_v=\alpha vPT\beta$$

and combining with (1.9)

$$\alpha=P\kappa\beta\Rightarrow P\beta=\frac{\alpha}{\kappa}\Rightarrow$$
$$c_p-c_v=\frac{\alpha^2vT}{\kappa} \tag{4.9}$$

Chapter 5

Statistical Thermodynamics and Third Law

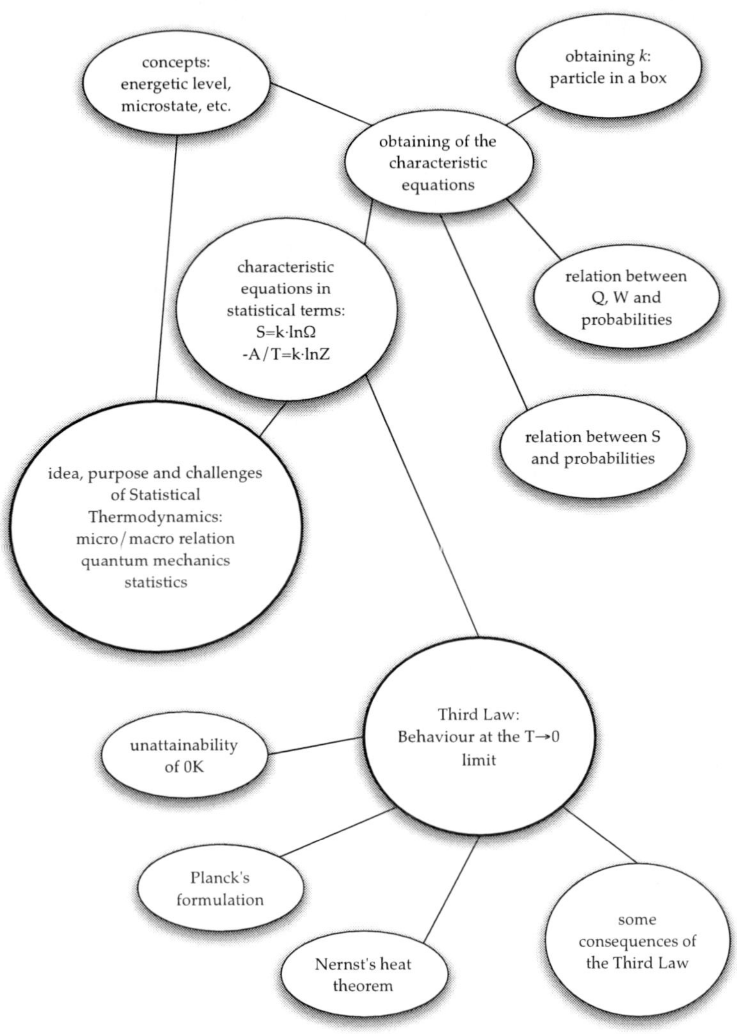

As we have seen, the minimal information required to describe a system thermodynamically is either c_p or c_v and its thermal equation. These allow reconstructing of the two branches of the derivation tree of a potential. However, thermal equation and specifc heat *must* be known. In engineering practice, they are read from tables of properties that summarize large series of measurements in laboratory.

By focusing on one particle of a certain substance and applying mechanics, one could progressively conceive how a number of them would interact; however big the number was. If one was suffciently brilliant, he could deduce from this the thermal equation of the substance, its specifc heat or any other property. This is exactly the way of statistical thermodynamics: to model the macroscopic behavior of a substance from the microscopic analysis of its particles, and then, deduce.

The process is not straightforward because it is not easy to know how micro and macro correspond, because the physics that normally explains the macroscopic is useless in explaining the microscopic. Since the number of particles involved is so huge, one can never be sure of anything; thus, being a question of probability.

This chapter gives a brief overview of statistical thermodynamics which will point out the major diffculties of the subject and the main ideas involved. However, the student is intended only to enjoy the conceptual structure as it unfolds within the chapter and hopefully illustrates many features about our usual macroscopic variables S, Q, and W, among others, which otherwise tend to remain obscure.

5.1. Goal of Statistical Thermodynamics

Goal. Statistical Thermodynamics is used to estimate the value of the thermodynamic properties of the systems, on the basis of knowledge about the structure and behavior of matter at microscopical scale.

Problems to be solved:

- *Connection between macroscopic and microscopic notions.* Basically two problems exist: (1) How can we conceive an irreversible process in 'microscopic' terms? (2) How are the macroscopic properties of a system related to its microscopic ones?
- *Physics associated with particles.* The nature and dynamics of particles in which substances are structured make that Quantum Mechanics must be taken into account to correctly describe their behavior.
- *Mathematical treatment:* Any macroscopic system we can fnd is formed by a number of particles so huge that it is necessary to assume simplifying hypothesis and to adopt statistical methodologies.

A complete analysis of a system through Statistical Thermodynamics would require:

1. To fnd a general method to relate the thermodynamic properties of the substance to its microscopic structure. The best way of doing that is through characteristic equations.

2. To validate the model with experimental evidence.

We only are going to see here step 1.

5.2. Characteristic Equations

The system's characteristic equation, as a function of its microscopic structure, is given by any of the following relations:

$$\text{Microcanonical ensemble:} \quad \boxed{S = k \ln \Omega} \tag{5.1}$$

$$\text{Canonical ensemble:} \quad \boxed{-\frac{A}{T} = k \ln z} \tag{5.2}$$

where $\Omega(U,V)$ is the number of quantum states that can be accessed by the system for the considered energy U and volume V, and $z(V,T)$ is the canonical partition function of the system, depending on system's volume and temperature T. To conveniently understand these concepts, we previously must study de Gibbs ensembles and the formulation of Thermodynamic Laws in microscopic terms.

5.3. Fundamentals

In Statistical Thermodynamics, we must describe systems at the particle scale (molecules, atoms, etc.).

> DEFINITION: A system's *microstate* is each of the possible quantum conf guration of that system.

> DEFINITION: *Energy level* is each of the possible values of the system's energy in a stationary quantum state.

> DEFINITION: *Energy level degeneration* is the number of microstates corresponding to a energy level.

The concepts of energy level and degeneration will be illustrated later through the study of a simple practical case. By this example the exposition of the method will be completed.

5.3.1. The First Law

In Statistical Thermodynamics, the First Law is a legacy of Mechanics. It is assumed that particles are mechanical systems. In a classical description, the mechanical state of the system would be completely def ned by the set of coordinates and momenta of all particles. As Classical Mechanics is not valid in a microscopic context, the microstate of the system is def ned by the equivalent information in Quantum Mechanics; that is the *wave function*. The wave function in Quantum Mechanics is obtained solving the fundamental Schrödinger equation. Since we are only interested in the description of equilibrium states, we focus on microstates with constant energy, called stationary quantum states. The wave function of these particular quantum states are the solutions of the stationary form of Schrödinger's equation:

$$\mathbf{H}\psi = E\psi$$

with **H** the operator corresponding to the total energy of the system, E the value of energy and ψ a function of the coordinates. Quantum Mechanics show that $|\psi|^2$ is the probability density of f nding all particles at each specif c coordinates. The wave function ψ biunivocally def nes a microstate. We can f nd two or more different wave functions that are solution of the Schrödinger equation for the same particular value of energy (degeneration).

5.3.2. Gibbs Ensembles and the Ergodic Conjecture

The value of the macroscopic mechanical variables (e.g., pressure), are considered to be the average in time of the values for the corresponding microscopic magnitude in the different microstates through which the system goes by.

However, it is very diff cult to treat this kind of averaging, since it would be a problem of dynamical evolution, not static. For this purpose, the tool of Gibbs ensembles is used:

> DEFINITION: A very big number $\mathcal{N} \to \infty$ of replicas of a system, all of them subject to the same macroscopic restrictions as the original system is said to be a Gibbs ensemble of that system.

The more interesting Gibbs ensembles are:

Microcanonical: representative of an isolated system with energy E.

Canonical: representative of a closed system in thermal equilibrium with a reservoir at temperature T.

Grand canonical: representative of an open system in thermal equilibrium with a reservoir at temperature T and in diffusive equilibrium with a particles source with chemical potentials μ_i for every different kind of particles i.

In Thermodynamics, to consider one or the other Gibbs ensemble for the subsequent analysis is merely a matter of convenience. Actually, in the microcanonical ensemble variables E and N (total number of particles) are f xed; in the canonical ensemble T and N are f xed; and in the grand canonical ensemble T and all μ_i are f xed. In any case, once these variables are f xed together with the system's volume V, the state of the system is def ned, and, consequently, all the thermodynamic functions too. The value of those functions does not depend on the variables chosen as the f xed ones. Hence, the selection of the Gibbs ensemble is equivalent to the choice of a particular set of independent variables. All three possible descriptions must lead to results with the same physical meaning.

Here, only the microcanonical (the easiest one from the point of view of the basic hypothesis formulation) and the canonical (the most useful one) ensembles are going to be considered.

We assume that *the value of the macroscopic mechanical variables coincides with the instantaneous average along the Gibbs ensemble representative of the system*. Therefore, if X is a macroscopic mechanical variable of a system in equilibrium:

$$\boxed{X = \bar{X} = \sum_r p_r X_r}$$

Where p_r is the probability of f nding one system of the ensemble, randomly chosen, in the particular quantum state r.

This statement is known as the *ergodic conjecture*. We do not give any proof of this because, in fact, it has been proved not to be true. Nevertheless, regarding the results that can be derived from it, it works; so we accept it.

5.3.3. The Second Law

As Boltzmann noticed, in a microscopical scale irreversible processes do not exist. Instead there exist processes whose inverse is extremely improbable.

It is well known that mechanics alone does not allow us to formulate Thermodynamics. Another another postulate, equivalent to the Second Law, must be added. In a microscopic scale, the Second Law is stated as follows:

Equal probability postulate: *A given isolated system in equilibrium can be found with equal probability in each of its accessible microstates*

This postulate leads immediately to the probability of a quantum state in the microcanonical ensemble: the isolated system under study has a f xed N, V, and E. Naming the number of steady quantum states accessible to the system in these conditions $\Omega = \Omega(N,E,V)$, and taking into account that the overall sum of probabilities must be 1,

$$\boxed{p_r = \frac{1}{\Omega}} \tag{5.3}$$

More precisely, Ω should be the number of quantum states with energy in the interval between E and $E + \delta E$, where δE is a small value in a "macroscopic scale", but it is big in a "microscopic scale"; at least in the order of a quantum step. The reason for this is the impossibility to determine, in a f nite time, the energy with complete precision due to Heisenberg's indetermination principle.

5.4. The Canonical Ensemble

DEFINITION: The so-called canonical partition function z is def ned as

$$\boxed{z = \sum_r e^{-\frac{E_r}{kT}}} \tag{5.4}$$

where $k = R/N_A \approx 1.38 \times 10^{-23}$ J/K is the Boltzmann constant.

The probability of a quantum state r in the canonical ensemble is given by

$$\boxed{p_r = \frac{e^{-\frac{E_r}{kT}}}{Z}} \tag{5.5}$$

The system's entropy is

$$\boxed{S = -k\sum_r p_r \ln p_r} \tag{5.6}$$

In order to proof them, we will use the probability of a quantum state in the micro-canonical ensemble (5.3). Additionally, we will need to analyze a simple case. Let us proof f rst relation (5.5)

PROOF

The canonical ensemble is formed by a number of closed systems that tends to inf nity. Each has N particles, volume V and f xed temperature T; however E may f uctuate.

This case is equivalent to an isolated ensemble of replicas, in which each of them is in thermal contact with its neighbors through diathermal walls. From the point of view of any of these replicas, the rest of the ensemble is seen as a reservoir at temperature T, since it has inf nite heat capacity and constant volume.

The ensemble (supersystem) satisf es:

1. Number of systems, $\mathcal{N}$: Systems cannot appear nor disappear. Let n_r be the number of systems in the quantum state r:

$$\mathcal{N} = \sum_r n_r \tag{5.7}$$

2. Energy, E_t: The supersystem is isolated, so its energy must be constant. If E_r is the energy corresponding to a system in the microstate r:

$$E_t = \sum_r n_r E_r \tag{5.8}$$

The previous points constrain the possible quantum states that a given system may be in, because it can only f nd itself in ones that make (5.8) hold for the supersystem. So, which states can the supersystem be in? It can be in all the possible combinations of the systems' (replicas) quantum states that satisfy (5.7) and (5.8).

DEFINITION: Every set of values $\{n_r\}$ is a distribution.

It does not matter which specif c systems are in a particular state r; the only thing that matters is how many are in each particular r state (for all the possible values of r).

Then, for a given distribution $\{n_r\}$, there are as many possible supersystem's states as different ways there are to assign the quantum states to the systems, without altering $\{n_r\}$.

1	2	1	2
2	2	2	2
3	4	1	4
4	2	3	4
5	5	2	1

Example: Canonic ensemble with $\mathcal{N} = 20$ systems. The quantum state of every system is indicated in the f gure.

Distribution: $n_1 = 4, n_2 = 8, n_3 = 2, n_4 = 4, n_5 = 2 \Rightarrow \sum n_r = 20 = \mathcal{N}$

If we swap the quantum states of the system in the lower left corner (5) and the system in the upper right corner (2), we get a different supersystem state but the same distribution.

Since every supersystem state is equally probable, according to the *equal probability postulate*, the most probable distribution is the one that contains the greatest number of combinations (supersystem different states). The number of the supersystem's quantum states for a given distribution $d = \{n_1, n_2, ...\}$ is the number of permutations with repetition of the $\mathcal{N}$ systems:

$$C_d = \frac{\mathcal{N}!}{\prod_r (n_r!)}$$

Maximizing the function $C_d = C_d(\{n_r\})$ with the restrictions (5.7) and (5.8), the set of values $\{n_r\}$ of the most probable distribution is obtained. We call it $\{n_r^*\}$, resulting:

$$n_r^* = \mathsf{A}e^{-\beta E_r}, \quad \forall r$$

It can be noticed that there are two parameters A and β, unique for all values of r. The f rst one is determined by considering the restriction (5.7): $\mathsf{A} = \mathcal{N} / \sum_r e^{-\beta E_r}$. We shall see later that $\beta = \frac{1}{kT}$.

n_r^* can be derived by means of the Lagrange multipliers method. In addition, Stirling's approximation must be applied: $\ln x! \approx x \ln x - x$.

For a given distribution, the probability of f nding any of the systems in a particular quantum state r is the fraction of systems with that quantum state, $n_r/\mathcal{N}$. It can be shown that the most probable distribution $\{n_r^*\}$ is much more probable than any other (≈ 1). Therefore we can assume that the probability of a quantum state is the one given by this distribution:

$$p_r = \frac{n_r^*}{\mathcal{N}} = \frac{e^{-\beta E_r}}{\sum_r e^{-\beta E_r}} = \frac{e^{-\beta E_r}}{Z} \tag{5.9}$$

5.4.1. Relation between Microscopical Variables and Q^R, W^R

Let us show that

$$dW^R = -\sum_r p_r dE_r \quad ; \quad dQ^R = \sum_r E_r dp_r$$

where dW^R and dQ^R are the work and heat respectively, in a elementary reversible process.

PROOF

For a system at rest and with no variation of potential energy, the internal energy (which in fact is a mechanical variable) will match with the average energy, accordingly with the ergodic conjecture:

$$dU = d\left(\sum_r p_r E_r\right) = \sum_r p_r dE_r + \sum_r E_r dp_r$$

Since E_r depends only on the metric variables X_i, and the generalized forces y_i are likewise the average of the values corresponding to the different quantum states $y_{i,r}$. We get:

$$y_{i,r} = -\left(\frac{\partial E_r}{\partial X_i}\right)_{X_{j\neq i}} \quad ; \quad y_i = \sum_r y_{i,r} p_r \tag{5.10}$$

$$\Rightarrow dW^R = \sum_i \left(\sum_r y_{i,r} p_r\right) dX_i \quad ; \quad dE_r = \sum_i \left(\frac{\partial E_r}{\partial X_i}\right)_{X_{j\neq i}} dX_i \Rightarrow$$

$$\sum_r p_r dE_r = \sum_r \sum_i p_r \left(\frac{\partial E_r}{\partial X_i}\right)_{X_{j\neq i}} dX_i = -\sum_i \sum_r p_r y_{i,r} dX_i = -\sum_i y_i dX_i =$$

$$= -dW^R \Rightarrow dU = \sum_r E_r dp_r - dW^R \Rightarrow dQ^R = T\,dS = \sum_r E_r dp_r$$

In other words: for a reversible process, the mechanical work modif es the quantum energy levels and does not modify the probabilities of the quantum states, while the heat exchange modif es the probabilities and does not modify the energy levels.

5.4.2. Relation between Probabilities and S

Let us demonstrate relation (5.6).

PROOF

Isolating E_r in (5.9)

$$E_r = -\frac{1}{\beta}(\ln p_r + \ln z) \Rightarrow T\,dS = -\frac{1}{\beta}\sum_r (\ln p_r dp_r) - \frac{1}{\beta}\ln z \sum_r dp_r$$

Since $\sum_r p_r = 1$, obviously $\sum_r dp_r = 0$, so

$$dS = -\frac{1}{\beta T} d\left(\sum_r p_r \ln p_r\right) \tag{5.11}$$

Because S is a state function, the second member must be an exact differential, so then

$$\frac{1}{\beta T} = f\left(-\sum_r p_r \ln p_r\right) = f(x) \Rightarrow S = \int f(x)dx = \varphi(x)$$

Additivity must be satisfed: $S(x_A)+S(x_B)=\varphi(x_A)+\varphi(x_B)=S(x_{A+B})=\varphi(x_{A+B})$

$$\begin{aligned}-x_{A+B} &= \sum_{r_{A+B}} p_{r_{A+B}} \ln p_{r_{A+B}} = \sum_{r_A, r_B} p_{r_A} p_{r_B} (\ln p_{r_A} + \ln p_{r_B}) = \\ &= \sum_{r_A} p_{r_A} \sum_{r_B} p_{r_B} \ln p_{r_B} + \sum_{r_B} p_{r_B} \sum_{r_A} p_{r_A} \ln p_{r_A} = \\ &= \sum_{r_B} p_{r_B} \ln p_{r_B} + \sum_{r_A} p_{r_A} \ln p_{r_A} = -(x_A + x_B)\end{aligned}$$

since A's quantum state is independent of B's one. Therefore

$$\varphi(x_A+x_B)=\varphi(x_A)+\varphi(x_B)$$

The solution of this mathematical equation is $\varphi = kx + \mathsf{C}$, where k is a universal constant and C is a constant that may vary from one system to another one, and which is additive (the constant of a compound system must be the sum of all the subsystem's constants: $\mathsf{C}_{A+B} = \mathsf{C}_A + \mathsf{C}_B$). Obviously, the origin of entropy depends on the value of C. By convenience, we establish $\mathsf{C} = 0$ for all systems, satisfying the additivity condition. In the section devoted to the Third Law we will discuss the implications of this. Finally,

$$S=-k\sum_r p_r \ln p_r \quad ; \quad \beta=\frac{1}{kT}$$

that demonstrates (5.5) from (5.9) as well.

It may be of interest to remark that the value of the universal constant k is not determined yet. However, is the same for any system, so we shall fnd it through the analysis of a particular case (the ideal gas) and it will hold for all.

5.4.3. Helmholtz Function

In section 5.2. we stated

$$A=-kT\ln z$$

PROOF

Isolating $\ln z$ in (5.5)

$$\ln z = -\frac{E_r}{kT} - \ln p_r \Rightarrow p_r \ln z = -\frac{p_r E_r}{kT} - p_r \ln p_r$$

The summation along all quantum states gives then

$$\sum_r p_r \ln z = \ln z = -\frac{\sum_r p_r E_r}{kT} - \sum_r p_r \ln p_r = -\frac{U}{kT} + \frac{S}{k} = -\frac{A}{kT} \Rightarrow$$

$$A=-kT\ln z$$

Notice that, regarding the partition function's defnition and the fact that the energy levels depend on the metric variables, $z = z(T, X_i)$. The expression (5.2) gives $A = A(T, X_i)$, which is the system's characteristic equation.

We may observe then that any thermodynamic function may be calculated if the system's partition function is known. Making use of the Gibbs-Helmholtz relation (for the particular case of a system with only one metric variable V),

$$-\frac{U}{T^2} = \left(\frac{\partial \frac{A}{T}}{\partial T}\right)_V = -k\left(\frac{\partial \ln z}{\partial T}\right)_V \Rightarrow U = kT^2\left(\frac{\partial \ln z}{\partial T}\right)_V \tag{5.12}$$

$$S = k \ln z + \frac{U}{T} \tag{5.13}$$

$$P = -\left(\frac{\partial A}{\partial V}\right)_T = kT\left(\frac{\partial \ln z}{\partial V}\right)_T \tag{5.14}$$

Obviously, for the validity of the previous relations the origins of U and S are no more arbitrary; contrarily they must be: the same one taken for the quantum states' energy for the internal energy, and the one corresponding to equation (5.6) for entropy.

5.4.4. The Ideal Gas. Determination of k

The value of the universal constant k can be obtained easily if we consider an ideal gas:

$$\boxed{k = \frac{R}{N_A} \qquad N_A = \text{Avogadro's number}} \tag{5.15}$$

The easiest system that we can apply Statistical Thermodynamics to, in order to calculate k, is the ideal gas.

In general, there is a term in the system's global amount of energy due to the potential energy arising from the many interactions between the different particles. If the density is low enough, those particles that are far from each other make the interactions between them weak enough to consider the potential energy negligible. In this case, the system's energy is reduced to the sum of the individual particles' energy (that is what we call *system of independent particles*).

Let N_s be the number of particles with mono-particle quantum state s, and ε_s the energy of that quantum state,

$$E_r = \sum_s N_s \varepsilon_s \tag{5.16}$$

The total energy of each particle can be split in the following way

$$\varepsilon = \varepsilon_{tr} + \varepsilon_{int} \tag{5.17}$$

where ε_{tr} is the translation energy of the particle, and ε_{int} is the energy associated with its internal degrees of freedom (rotation, vibration, and so on). We shall see in section 5.4.5. that ε_{tr} depends on the system's metrical variables. On the other hand, ε_{int} does not depend on the metrical variables, since it is related to movements relative to the particle's center of gravity.

Let us consider a pure gas formed by independent, non-distinguishable particles.

PROOF

From (5.16) and (5.17), we can deduce $Z = Z_{\text{tr}} Z_{\text{int}}$.

Quantum Mechanics (see section 5.4.5.) provides that the energy of translation along axis x of one free particle inside a box of length L_x is $\varepsilon_{\text{tr},x} = \frac{\text{h}^2}{8m}\frac{\text{n}_x^2}{L_x^2}$, where n_x is a positive integer quantum number. Hence, the *canonical partition function* of one free particle with movement along axis x is given by:

$$Z_{1,\text{tr},x} = \sum_{\text{n}_x=1}^{\infty} \exp\left(-\frac{\text{h}^2}{8mkT}\frac{\text{n}_x^2}{L_x^2}\right) = \frac{L_x}{\text{h}}\sqrt{2\pi mkT}$$

This result was achieved by approximating the summation by an integral (then, in this case, not valid for very low temperatures). Considering the three axis directions:

$$\varepsilon_{\text{tr}} = \varepsilon_{\text{tr},x} + \varepsilon_{\text{tr},y} + \varepsilon_{\text{tr},z}$$

$$Z_{1,\text{tr}} = Z_{1,\text{tr},x} Z_{1,\text{tr},y} Z_{1,\text{tr},z} = V\left(\frac{2\pi mkT}{\text{h}^2}\right)^{\frac{3}{2}}$$

Now, if we consider N distinguishable particle inside the box, the swap of the quantum states of two particles will bring a different quantum state of the system. $E_{\text{tr}} = \sum_{n=1}^{N} \varepsilon_{\text{tr},n}$ leads to $Z_{\text{tr}} = (Z_{1,\text{tr}})^N$.

To illustrate this, let us suppose two distinguishable particles A and B:

$$Z_{\text{tr}AB} = \sum_{\text{n}_{AB}} e^{-\frac{E_{\text{tr}AB}}{kT}} = \sum_{\text{n}_A,\text{n}_B} e^{-\frac{\varepsilon_{\text{tr}A}+\varepsilon_{\text{tr}B}}{kT}} = \sum_{\text{n}_A} e^{-\frac{\varepsilon_{\text{tr}A}}{kT}} \sum_{\text{n}_B} e^{-\frac{\varepsilon_{\text{tr}B}}{kT}} = Z_{\text{tr}A} Z_{\text{tr}B} = Z_{1,\text{tr}}^2$$

Nevertheless, we are dealing with a pure ideal gas, and the particles are non-distinguishable. Permutation of the particles' quantum states **does not** produce a new system's quantum state hence. By the previous calculation, we would have just counted each system's quantum state a number of times $N!$ The correct partition function is

$$Z_{\text{tr}} = \frac{1}{N!} V^N \left(\frac{2\pi mkT}{\text{h}^2}\right)^{\frac{3N}{2}} \tag{5.18}$$

This expression is only valid if the number of microstates accessible to the particles is much greater than the number of particles (since statistical reasoning associated with the symmetrization postulate of Quantum Mechanics has not been considered). This condition is satisf ed in almost any system that is of interest for an engineer.

The thermal equation of state resulting from the connection between P and Z, noticing that Z_{int} does not depend on V, is:

$$P = kT\left(\frac{\partial \ln Z_{\text{tr}}}{\partial V}\right)_T = \frac{NkT}{V} \Longrightarrow PV = NkT$$

Identifying this with the thermal equation of state for an ideal gas $PV = \frac{N}{N_{\text{A}}}RT$, we get

$$k = \frac{R}{N_{\text{A}}} = \frac{8.314472(15)\ \text{J/(mol·K)}}{6.022\,141\,79(30) \times 10^{23}\ \text{mol}^{-1}} = 1.380\,6504(24) \times 10^{-23}\ \text{J/K}$$

5.4.5. Particle in a Box

Imagine one particle inside a box (see next f gure) of dimensions L_x, L_y, and L_z, with one vertex placed on the coordinate system origin and its edges matching the axis.

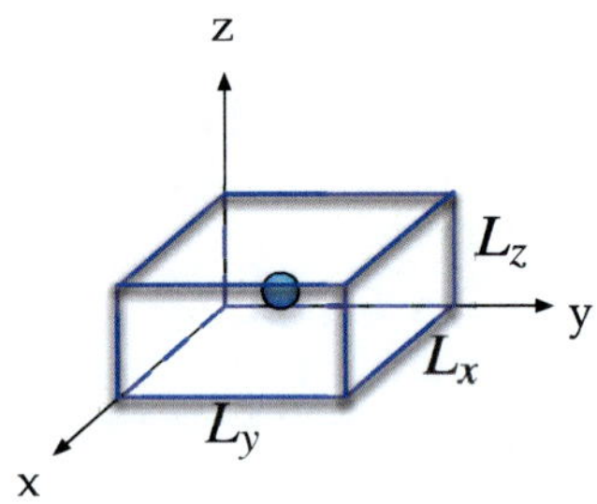

The *translation energy* of the particle is:

$$\varepsilon_{tr} = \frac{h^2}{8m}\left(\frac{n_x^2}{L_x^2} + \frac{n_y^2}{L_y^2} + \frac{n_z^2}{L_z^2}\right) \qquad n_x, n_y, n_z = 1, 2 \ldots \tag{5.19}$$

with $h = 6.626 \times 10^{-34}$ J·s the Planck's constant; m: particle's mass.

The solutions of the stationary Schrödinger's equation are:

$$\psi(x,y,z) = \sqrt{\frac{8}{V}} \sin\frac{n_x \pi x}{L_x} \sin\frac{n_y \pi y}{L_y} \sin\frac{n_z \pi z}{L_z} \tag{5.20}$$

Notice that every system's microstate becomes f xed by the metrical variables L_x, L_y, L_z, and the *quantum numbers* n_x, n_y, n_z, which may adopt any positive integer value.

PROOF

The previous expressions can be derived from the application of the stationary Schrödinger's equation to a free particle (null potential energy) that moves unidimensionally on a line segment between the positions $x = 0$ and $x = L_x$. Calling the particle energy ε_x, the equation to be solved is

$$-\frac{h^2}{8\pi^2 m}\frac{\partial^2 \psi_x}{\partial x^2} = \varepsilon_x \psi_x$$

which gives

$$\psi_x = A \sin\frac{x\sqrt{8\pi^2 m \varepsilon_x}}{h} + B \cos\frac{x\sqrt{8\pi^2 m \varepsilon_x}}{h}$$

In order to determine constants A and B, the following conditions are imposed: (1) ψ_x must be continuous and null outside the segment $(0, L_x)$ (since the probability of f nding the particle out of the box vanishes), and hence $\psi(0) = 0$, $\psi(L_x) = 0$; (2) the normalization condition $\int_0^{L_x} \psi^2 = 1$. From this boundary conditions:

$$B = 0 \quad ; \quad A = \sqrt{\frac{2}{L_x}} \quad ; \quad \varepsilon_x = \frac{n_x^2 h^2}{8mL_x^2}$$

$$\psi_x(x) = \sqrt{\frac{2}{L_x}} \sin \frac{n_x \pi x}{L_x}$$

To extend the result to the whole three-dimensional domain, it suff ces to multiply the analogously obtained for the directions y and z:

$$\psi(x,y,z) = \psi_x(x)\psi_y(y)\psi_z(z)$$

and to add the three components' energy:

$$\varepsilon_{\text{tr}} = \varepsilon_x + \varepsilon_y + \varepsilon_z$$

In this case, only the three quantum numbers related to the particle's position n_x, n_y, n_z take part. More complex systems would require a larger number of state variables (quantum numbers) to quantify rotation, vibration, electronic movement, magnetic energy, spin, etc., for *all* the present particles, and even more quantum numbers would be required to describe the particles' interactions.

As we have just seen, the possible energy values for the system are some specif c values; every possible value of energy is called *level*. The energy level with the lowest value (in the case just studied, that would arise for the three quantum numbers equal to 1) is named *fundamental level*. Several microstates (different wave functions ψ) may have the same value of energy (e.g., in the particle in a box case with $L_x = L_y = L_z$, the three quantum states with two quantum numbers equals to 1 and the other one equal to 2). The number of different microstates corresponding to the same energy level is called *level degeneration*.

5.5. Entropy and Ω. The Boltzmann's Principle

Let us prove the characteristic equation (5.1), known as the Boltzmann's principle.

PROOF

Our starting point is the relation between entropy and probabilities obtained for the canonical ensemble

$$S = -k \sum_r p_r \ln p_r$$

1. Let p_E be the probability of the system having energy E in the canonical ensemble, and p_{r_E} the probability of the system that has energy E being in the quantum state r_E. We immediately get $p_r = p_E p_{r_E}$.

2. Expanding the S expression:

$$S=-k\sum_{E,r_E} p_E p_{r_E}\ln(p_E p_{r_E}) = -k\left(\sum_{E,r_E} p_E p_{r_E}\ln p_E + \sum_{E,r_E} p_E p_{r_E}\ln p_{r_E}\right)$$

$$\sum_{E,r_E} p_E p_{r_E}\ln p_E = \sum_{E}\left(p_E \ln p_E \sum_{r_E} p_{r_E}\right) = \sum_{E} p_E \ln p_E$$

3. It is possible to show that the energy f uctuations in a stable macroscopic system are negligible. If the average energy is $\overline{E}$, the range in which the probability is not practically null is of the order of $\Delta E/\overline{E} \sim 1/\sqrt{N}$, where N is the number of particles in the system. Therefore, for systems with a large number of particles (of the order of the Avogadro's number for the macroscopic usual systems), we can assume that $p_E \approx 1$ for $E = \overline{E}$ and $p_E \approx 0$ in the rest of the cases. It is concluded $\sum_E p_E \ln p_E \approx 0$

4. Let Ω be the number of quantum states with energy $\overline{E}$ (more precisely, with energy between $\overline{E} - \Delta E/2$ and $\overline{E} - \Delta E/2$). The probability of all these microstates is the same: $p_{r_{\overline{E}}} = \frac{1}{\Omega}$. Then

$$S=-k\sum_{r_{\overline{E}}} p_{r_{\overline{E}}}\ln p_{r_{\overline{E}}} = -k\sum_{j=1}^{\Omega}\frac{1}{\Omega}\ln\frac{1}{\Omega} = k\ln\Omega$$

5.6. The Third Law

The so-called Third Law is not really a law, but some statements related to the behavior for systems near absolute zero. It may seem that this behavior is not interesting for engineers, as they very rarely f nd a system in such situations. But the fact is that it has a lot of applications, for example, especially in reacting systems. Furthermore, at temperatures close to zero, things like superf uidity or superconductivity appear; they may have, in a not too distant future, interesting engineering applications, if systems that exhibit the same behavior at reasonable temperatures are found.

Behavior in the limit 2 *(part 1)* ***Unattainability of* 0 K**
It is impossible for any system and by any procedure to reach absolute zero in a f nite number of steps

Behavior in the limit 2 *(part 2)* ***Hypothesis of Planck***
The entropy of any pure, perfectly crystalline substance approaches zero as temperature approaches absolute zero

Behavior in the limit 2 *(part 3)* ***Nernst's heat theorem***
The entropy change approaches zero for any reversible isothermal process that occurs between two equilibrium states as the temperature approaches absolute zero

5.6.1. The Unattainability of Absolute Zero Theorem

Liquid helium, whose standard boiling point is 4.2 K, can be got through criostates, extremely well isolated containers. If this helium is evaporated at low pressures, temperatures down to 0.7 K can be obtained. To reach lower temperatures, it is necessary to use successive magnetization and demagnetization of paramagnetic substances (magnetizable), as chromium alum. With this method, a magnetic f eld that lines up the electronic spins is used, and the material is then cooled in a liquid helium bath. When the magnetic f eld disappears, the spins adopt again a random orientation, taking energy from the particles' movement, what makes temperature decrease. Demagnetization of paramagnetic salts leads to temperatures of only 2 mK, and demagnetization of atomic nuclei allows to reach temperatures of only 10^{-5} K. For other more complex systems, temperatures even lower have been attained, reaching less than 1 nK.

Experience shows that the colder a system is, the harder is to cool it further; resulting that the absolute zero is, in practice, unattainable. This experimental conclusion cannot be justif ed from the First or Second Laws of Thermodynamics.

5.6.2. Planck's Formulation

As seen, $p_r = \frac{1}{z}\exp\left(\frac{-E_r}{kT}\right)$, so when T decreases, the probabilities of the quantum states with the largest energy decrease too. Furthermore, at $T = 0\,\mathrm{K}$ the only energy level that could be occupied is the fundamental one.

The probability of a certain energy level E_i (instead of that of a particular quantum state r: p_r, that we had been using until now) is:

$$p_i = \frac{g_i e^{-\frac{E_i}{kT}}}{\sum_i g_i e^{-\frac{E_i}{kT}}}$$

where g_i is the i-level degeneration. So, when approaching absolute zero, we have $\lim_{T\to 0} p_{i=0} = 1$ (fundamental level, E_0) and $\lim_{T\to 0} p_{i\neq 0} = 0$. In words: that the probability of any energy level other than the fundamental tends to zero.

If we now calculate the entropy of the system as we approach the absolute zero:

$$\lim_{T\to 0} S = -k \lim_{T\to 0} \sum_r p_r \ln p_r$$

Logically, the sum has g_0 terms, one for each of the possible states associated to E_0. The rest of them are zero, because their probability is zero and $\lim_{x\to 0} x \ln x = 0$. So, for the possible states:

$$\sum_r p_r = 1 \Rightarrow p_r = \frac{1}{g_0} \Rightarrow p_r \ln p_r = \frac{1}{g_0} \ln \frac{1}{g_0} \Rightarrow$$

$$\lim_{T\to 0} S = -k g_0 \left(\frac{1}{g_0} \ln \frac{1}{g_0}\right) = k \ln g_0$$

If the fundamental level is non-degenerated (there is only one possible state):

$$\lim_{T\to 0} S = k \ln 1 = 0$$

Planck's formulation is: The entropy of any pure, perfectly crystalline substance approaches zero as temperature approaches absolute zero.

This formulation would require to quantify the conf gurations corresponding to crystalline structure disorder, which will be addressed later, also including the different possible nuclear spin orientations and the isotope mix. In other words, the real absolute entropy is **not** null. Nevertheless, it is the generally accepted convention to not consider the last mentioned conf gurations, since as a matter of fact they tend to stay practically unaltered in any physical or chemical transformation that does not affect the atomic nucleus nor the isotope mix. Contrarily, the entropy due to the crystalline disorder cannot be deprecated; it originates a residual entropy over the zero conventional value. Regardless, in these particular cases, the entropy can be considered zero at null temperature in most of the systems, and *Planck's formulation can be accepted* as a *practical reference* to establish an "absolute" entropy scale from calorimetric data. The standard absolute entropy tables (from 0 K) already include the needed corrections to take into account the previous consideration.

5.6.3. Nernst Heat Theorem

Notice that, even if $\mathsf{g}_0 \neq 1$, it is still valid

$$\lim_{T\to 0} \Delta S_T = k \ln \frac{\mathsf{g}_0^2}{\mathsf{g}_0^1} = 0 \qquad (5.21)$$

while $\mathsf{g}_0^2 = \mathsf{g}_0^1$, as stated by Nernst heat theorem.

This theorem was previously known in an empiric way through a generalization of what was observed in a variety of processes (phase transitions, chemical reactions, etc.). The apparent contradictions, in most of the cases, are due to:

1. The phases are not crystalline. This may happen because, when cooling, the molecules have not been able to f nish reordering into a regular structure before 'freezing'. Then, it is said that the system *has not achieved a stable internal equilibrium*. The vitreous phases (liquids sub-cooled under their solidif cation temperature) are a typical example.

2. The crystals are formed by links between different atoms (CO, NO, NO_2, H_2O, etc.) whose molecular asymmetry allows non-regular ordinations and a consequent somehow disordered conf guration.

In such cases, Nernst's theorem is not satisf ed because, during the phase transition or the chemical reaction, molecular ordination is modif ed.

All exceptions to Nernst's heat theorem are also exception to Planck's hypothesis. It can be proved that the Nernst heat theorem is a suff cient condition (but not necessary) of the unattainability of absolute zero theorem.

5.6.4. Some Consequences of the Third Law

From (5.21) we immediately obtain

$$\lim_{T\to 0}\left(\frac{\partial S}{\partial P}\right)_T = -\lim_{T\to 0}\alpha V = 0 \Rightarrow \lim_{T\to 0}\alpha = 0 \tag{5.22}$$

$$\lim_{T\to 0}\left(\frac{\partial S}{\partial V}\right)_T = \lim_{T\to 0}\beta P = 0 \Rightarrow \lim_{T\to 0}\beta = 0 \tag{5.23}$$

and similar equations in electric systems, magnetic systems, etc., referred to the analogous coeff cients (electric or magnetic susceptibility, etc.).

Regarding heat capacity,

$$S(T,V) - S(0,V) = \int_0^T \left(\frac{\partial S}{\partial T}\right)_V dT = \int_0^T \frac{C_v}{T} dT \neq \infty \Rightarrow \boxed{\lim_{T\to 0} C_v = 0} \tag{5.24}$$

and similarly

$$\boxed{\lim_{T\to 0} C_p = 0} \tag{5.25}$$

From Nernst heat theorem, in phase changes taking place at $T \to 0$, the Clausius-Clapeyron equation (it will be seen later) indicates

$$\lim_{T\to 0}\frac{dP^{\text{sat}}}{dT} = \lim_{T\to 0}\frac{\Delta s^{\text{sat}}}{\Delta v^{\text{sat}}} = 0 \tag{5.26}$$

since $\lim_{T\to 0}\Delta v^{\text{sat}} \neq 0$. The slope of the saturation curve must be null at absolute zero.

All these consequences of the Third Law have been conf rmed experimentally in many cases (He solid-liquid equilibrium, Sn white-gray allotropic transformation, Debye's Law of crystalline solids specif c heat at low temperature $c_v \approx \mathsf{a}T^3$, etc.).

Chapter 6

Pure Substances

6.1. Introduction

Pure substances are an abstraction; they never exist in reality, because substances around us are always mixtures: mixtures of several chemical species and mixtures of the several isotopes for every element. However, in many cases, the proportions of several components remain f xed in all processes of interest (as it is usually the case with isotopes in the industry, except precisely when the aim is their separation), so that they can be considered as only one effective component. Moreover, sometimes a substance is mainly formed by one effective component and minor proportions of others, so that its overall behavior can be modeled as if it consisted of only the main component. This is what we shall normally call a *pure substance*.

In this chapter, we shall treat mainly true thermodynamic equilibrium states, which are sometimes denoted as stable equilibrium states. Nevertheless, we shall also consider metastable states, because they appear in practice from time to time.

6.2. Overview of the Behavior of Pure Substances

There are several distinct forms that a substance may adopt depending on the pressure and temperature at which it f nds itself. These are solids and f uid. Usually, f uids are said to be liquid, gas, or supercritical f uid; but, as we shall see, there is no strict distinction between them. Although all substances show these distinct forms, the values of pressure and temperature for which they adopt one or the other, and their values at which they change are specif c to each substance.

> DEFINITION: *Solid* is a state of matter characterized by having a crystalline structure at the microscopic level.

True solids are always crystals. As there are many possible arrangements of the particles (molecules, ions) to form a net (cubic, hexagonal, and so on), most substances can adopt several solid forms that are stable in different regions on a P-T diagram. Transitions from one solid form to another are f rst-order transitions, as all the transitions that we are going to study, so they don't differ in any important aspect from, for example, a liquid-solid transition.

Amorphous solids (sometimes called glasses, the typical example) are not true solids, but rather liquids with infinite viscosity. Glasses are metastable states, though they can last for centuries. Therefore, the transition between the amorphous solid and the liquid state (called *glass transition*) is not a first-order transition: the change at a given pressure does not happen at a fixed temperature, but in a certain temperature range.

DEFINITION: *Fluid* is the state of matter characterized by a basically random structure at the microscopic level.

In a fluid, particles are placed more or less at random (not completely in general, because intermolecular forces may introduce a certain non-randomness). Particles can move relatively to each other. When they are very closely packed, the fluid is usually called *liquid*, and when they are more loosely packed, *gas* or *vapor*.

Helium has two liquid forms, normal and superfluid. Normal liquid-superfluid liquid transition is not first-order; it is called a lambda-transition. Non-first order transition will not be studied here, as there is very little interest for engineering applications.

DEFINITION: The *binodal curve* is made from all the points that represent a single fluid phase in a two-fluid phases equilibrium state.

When the binodal curve is represented in a *P-T* diagram, it is called the *vapor pressure curve*.

The heterogeneous two-fluid phases equilibrium states at a given pressure (or at a given temperature) are always a mixture of the same two states; the denser one is called saturated liquid and the other, saturated gas.

DEFINITION: *Saturated liquid* is an homogeneous fluid that can be at equilibrium with another fluid phase (of the same pure substance) without any restriction to the interchange of mass or energy, and is the denser of both.

DEFINITION: *Saturated gas* is the other phase that may coexist with saturated liquid.

DEFINITION: *Critical point* is the state of the saturation line where the liquid branch meets the gas branch.

At the critical point, there is no liquid or gas, just a single fluid phase. The *critical properties* (critical temperature, critical pressure, and so on) are characteristics of the substance. Critical pressure is usually in the order of tens of bars, although it can be as great as 220 bar for water, or even more than 5 000 bar for gold.

DEFINITION: *Triple point* are the states of an heterogeneous pure substance with one solid phase and two fluids (liquid and gas).

It can be shown that a triple point can exist only at a given pressure and temperature. Its coordinates (P_t, T_t) are also characteristic of the substance. For instance, for water they are: 611 Pa, 273.16 K. Most substances have low triple pressure; carbon dioxide with a triple point at (526.89 kPa, 216.75 K) is the most signif cant exception.

> Helium is a interesting exception, in that it has no solid-liquid-gas triple point, although it has other triple points. For example, He^4 has a liquid II-solid (hexagonal)-solid (centered cubic) triple point at about 1.5 K, and He^3 has a liquid-solid (centered cube)-solid (hexagonal) and a liquid-solid (centered on faces cube)-solid (hexagonal). This means that for helium the vapor pressure goes down to $0\,K$, giving one of the experimental evidences for the Nernst Heat Theorem, as we saw in equation (5.26).

DEFINITION: *Sublimation curve* is the line of the *P*-*T* diagram where all the solid-gas equilibrium states projects.

The sublimation curve goes from the triple point downwards. In the *P*-*v*, *T*-*v*, or *PvT* diagrams, the sublimation zone is a surface.

DEFINITION: *Solidif cation curve* is the line of the liquid states that may coexist in equilibrium with a solid.

It is called so because if we cool at constant pressure a liquid in this curve, solid begins to form. It goes from the liquid in the triple point upwards.

DEFINITION: *Melting curve* is the line of the solid states that may coexist in equilibrium with a liquid.

It is called so because if we heat at constant pressure a solid in this curve, liquid begins to form. It goes from the solid in the triple point upwards. Solidif cation and melting curves are the same curve in a *P*-*T* diagram and, as can be shown from the Clausius-Clapeyron equation, that we will study later; this curve is almost a vertical line, that is, the melting temperature of a solid varies only very slightly with pressure.

DEFINITION: *Vapor quality* of a liquid-vapor system is the mass fraction of vapor.

Vapor quality is usually represented as x, and is an dimensionless number ranging from 0 (saturated liquid) to 1 (saturated vapor). All lines of constant quality go from the triple line to the critical point, where they meet. Vapor quality is frequently used for the def nition of the state of a liquid-vapor system, together with another thermodynamic property (usually P or T).

The state of other two-phases systems are usually identif ed in an analogous way, using the mass fraction of one of the phases x (the other phase mass fraction is necessarily $1-x$).

For the complete def nition of the state of a three-phase one-component system, two mass fractions are needed (for example, of solid, x^S, and liquid, x^L).

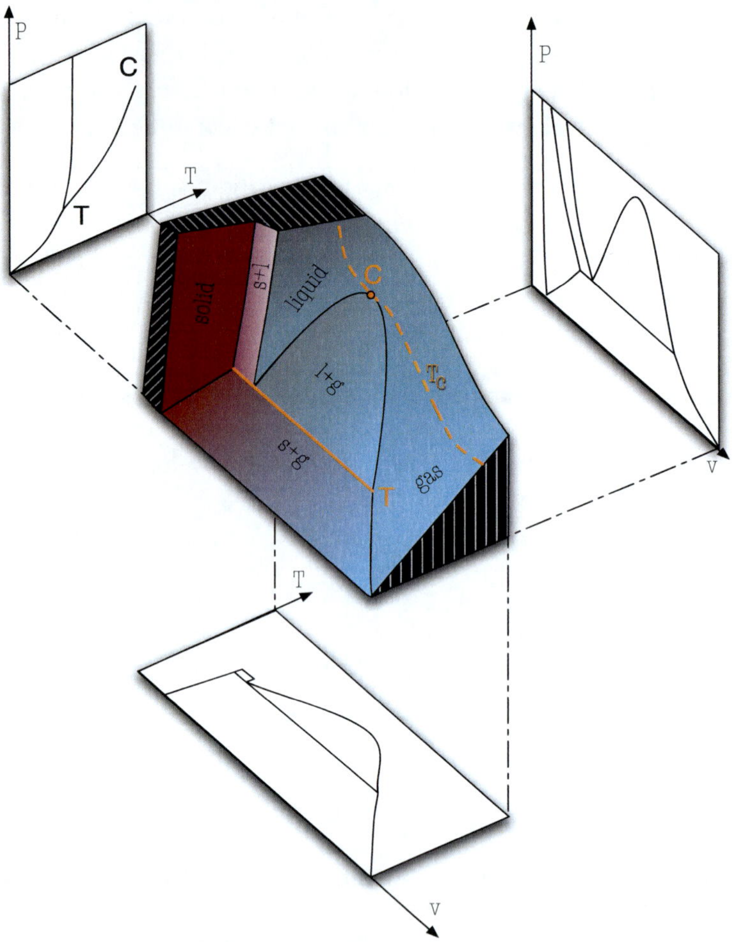

Figure 6.1. PvT surface and its projection on the three planes. Features of special interest in the surface are indicated in orange: critical point, critical isotherm, and triple line.

6.2.1. Description of the *PvT* Surface

> DEFINITION: The *PvT* surface of a pure substance is the graphical representation of its specif c volume for all the values of pressure and temperature of stable equilibrium states.

The three-dimensional surface can be projected on any of the three planes *P*-*T*, *P*-*v*, and *T*-*v*, leading to the corresponding two-dimensional diagrams. The surface and their projections are shown schematically in f gure 6.1. The *P*-*v* projection of the *PvT* surface can be seen in more detail in f gure 6.2.

Solid: Figure 6.1 labels as *solid* an area of the *PvT* surface which is represented as homogeneous. As it has been said before, most substances may adopt more than one form of solid conf guration. The area of the surface representing *solid* in a real diagram

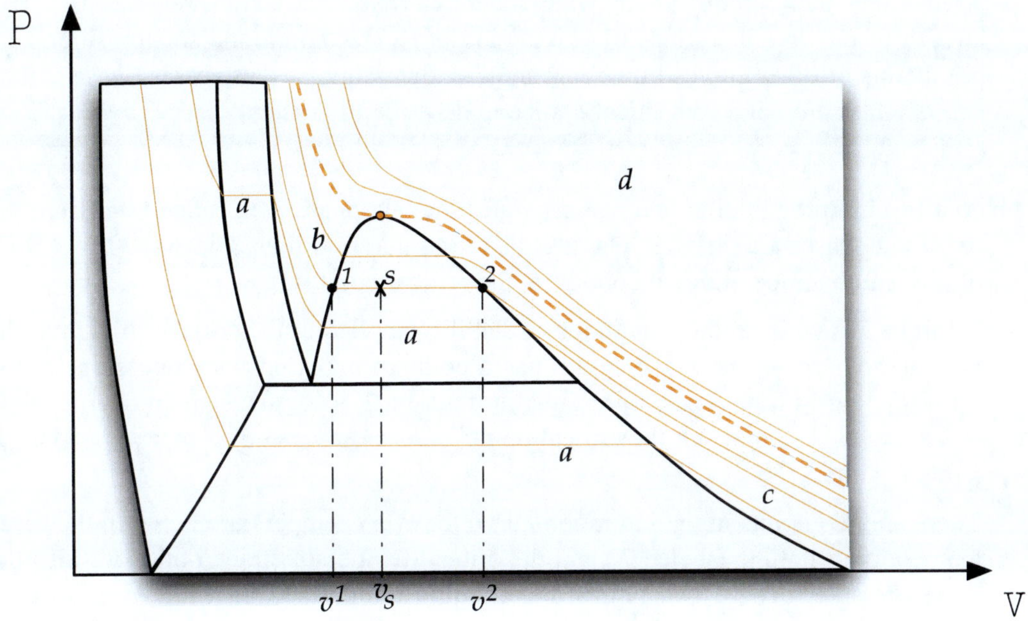

Figure 6.2. *P*-*v* projection of the *PvT* surface.

might accordingly be subdivided into regions representing the solid forms, with two-solid phases regions between them, and triple lines with three-solid phases or two solids and one liquid.

The specif c volume of a solid usually varies only a little with temperature and even less with pressure.

Fluid: It is the continuous surface that surround the area marked as '1+g' from above. The region labeled 'b' in f gure 6.2, which is limited by the dashed line representing the critical temperature T_c of the substance (which will be explained further on), and the 'V' given by the $s+l$ and $l+g$ areas is frequently called *liquid*. The area 'c', between T_c the $s+v$ and $l+g$ areas and $v=\infty$, is frequently called *gas* or vapor. The area of temperatures higher than T_c, is sometimes called *supercritical f uid* (area 'd').

The distinction between liquid and gas seems natural to us because we can appreciate clear differences between the properties of substances, such as water, when we see them in gaseous form (its density is low and varies noticeable with pressure) and in liquid form (its density is high and almost independent of pressure). Physically, however, there is no sharp limit between liquid and gas. In fact, one could pass from one to the other by an adequate process through the 'd' area, avoiding to cross the two-phase area, without any discontinuity in the evolution of the properties of the substance.

> Remember that certain liquids (like water) can have, at certain regions, $\alpha < 0$, which results in some kinds of 'caves' in the *PvT* diagram and superposition of isotherms or other curves in the *P*-*v* and *T*-*v* diagrams; in these cases, these pairs of variables are not a convenient choice to describe the system.

Saturated liquid and gas: the line that separates the one-phase area called f uid from the two-f uid phases area labeled *l*+*g* is called the saturation line. The curve has a point of maximum temperature and pressure, the critical point (C).

It can be shown from the conditions of stability applied at the critical point, that the critical isotherm in the *P*-*v* diagram has a point of inf exion with zero slope at the critical point, as it has been represented in f gure 6.2; also, that some properties tend to inf nity as we approach the critical point: $\alpha^c = \infty, \kappa^c = \infty, c_v^c = \infty, c_p^c = \infty$; but β^c remains f nite.

There are two points in the saturation curve for every temperature between the triple temperature (the line labeled *T*) and the critical temperature, and both have also the same pressure. These two points are in equilibrium; the one with lower specif c volume is called *saturated liquid* and the other is called *saturated gas*. In fact, states along the saturation lines (except the critical point) are the only states that can be called liquid or gas without ambiguity.

When the saturation line is represented in a *P*-*T* diagram, the two parts of the line superposed, and they appear as a single curve, called the *vapor pressure curve*. The vapor pressure curve of every substance goes from the triple point to the critical point.

The saturation line has a maximum at the critical point, both in a *T*-*v* diagram (the critical temperature), and in a *P*-*v* diagram (the critical pressure). At the other side, it ends in the triple line, at the points that represent respectively the gas and the liquid at the triple point.

Two-phase areas: Three areas in f gure 6.1 appear labeled as *s*+*g*, *s*+*l*, and *l*+*g*, which stand for solid+gas, solid+liquid, and liquid+gas. In the states contained in these areas, a given quantity of substance will appear in the corresponding two phases at the same time, a fraction of the total mass in one form and the rest in the other.

All the points in one of these surfaces along a given temperature have the same pressure, so they are project in a *P*-*T* diagram in a single point; the whole area is projected in a curve, the vapor pressure curve in the case of *l*+*g*, the *sublimation curve* in the case of *s*+*g* (from the triple point downwards), and the *melting* or *solidif cation* curve in the case of *s*+*l* (from the triple point upwards). Sublimation can therefore happen only at pressures under the triple pressure, and that's why so few substances sublimate at ambient pressure.

It is a characteristic of the liquid-gas transition that its two-phase area ends in a maximum (the critical point); all substances have a critical point, where the saturated liquid and saturated gas meet. Nothing similar happens with other transitions. The consequence is that there is no sharp distinction between liquid and gas, but there is between solid and f uid, and between different kinds of solids.

Three-phase line: It is the line labeled *T*, which represents the states in which the substance can have three different phases in equilibrium: solid, liquid, and gas. This can happen only at a given pressure and at a given temperature, characteristic of the substance (triple pressure and triple temperature); the specif c properties of each phase at the triple states are always the same, only the mass fractions in each phase can vary. The triple line projects in a single point in a *P-T* diagram, and in the corresponding lines in the *P-v* and *T-v* diagrams. To have one and only one three-phase state represented by a point, a *h-v* diagram can be used.

Substances may become reactive at high temperatures, decomposing or forming dimers. Then, the form of the *PvT* surface might be very different, as the system becomes multicomponent.

6.3. Thermodynamical Formulation of Heterogeneous States

An heterogeneous system in equilibrium must necessarily be in thermal and mechanical equilibria; so, temperature and pressure should be the same for all phases. Therefore, the pair of variables (P,T) is not a good choice to def ne these states. An alternative to choosing some specif c variable (but not g) is the mass fraction, usually called quality in the case of a vapor-liquid system.

6.3.1. Conditions for Multiphase Equilibria

Let us think about a certain quantity of substance contained in a closed, rigid and adiabatic recipient, and let us assume that in the state of equilibrium, the substance present two phases, 1 and 2. Three conditions must be met:

$$T^1 = T^2 \tag{6.1}$$

$$P^1 = P^2 \tag{6.2}$$

$$g^1 = g^2 \tag{6.3}$$

The f rst one is the condition of thermal equilibrium. The second is the mechanic equilibrium, provided that we neglect the effect of external f elds (like, for example, gravity) and the energy of the interface.

If the gravity f eld is taken into account, hydrostatic pressure will appear. Also, if the energy of the interface is not negligible, the surface tension will appear in the equation. Nevertheless, the f rst effect is important only if we consider a height of some decimeters at least of liquid, and the second, only with the very f rst bubbles formed in a liquid or very small droplets in a gas. So, for most engineering applications, they can be neglected.

The third condition is *diffusion equilibrium*: the specif c Gibbs function of both phases must be equal. Being so, the transfer of mass between both phases would be reversible and can take place in both directions, so the system is in equilibrium. In the next chapter, we shall see a more practical formulation in equation (7.10).

PROOF

For the sake of simplicity, we shall consider a closed system, without any restriction to transfer of mass or energy between the phases. We must derive f rst the form of the Gibbs equation for the system. We must bear in mind that mass is variable in each phase, extensive properties are additive, and P and T are the same in both phases. For a single phase (open system):

$$\begin{aligned} dU &= dG - d(PV) + d(ST) = g\,dn + n\,dg - P\,dV - V\,dP + S\,dT + T\,dS \\ &= g\,dn - \cancel{ns\,dT} + \cancel{nv\,dP} - P\,dV - \cancel{V\,dP} + \cancel{S\,dT} + T\,dS \Rightarrow \\ dS &= \frac{dU}{T} + \frac{P}{T}dV - \frac{g}{T}dn \end{aligned}$$

Now, for the closed system, every extensive property (S, U, V, n) is the sum of both phases. We must remember that a system is in thermodynamic equilibrium if its properties don't change with time when the system remains isolated. So, we must consider the system isolated, which means that $dU = dU^1 + dU^2 = 0$ (no energy change), $dV = dV^1 + dV^2 = 0$ (so that there will be no work), and $dn = dn^1 + dn^2 = 0$ (closed system). Thus, $dn^2 = -dn^1$. Finally, the entropy of the system cannot decrease, only increase or remain constant, because it is a closed and adiabatic system. But if the entropy should increase, the system would not had been in equilibrium, so $dS = 0$, therefore in equilibrium:

$$dS = 0 = dS^1 + dS^2 = \frac{\cancelto{0}{dU^1 + dU^2}}{T} + \frac{P}{T}\cancelto{0}{(dV^1 + dV^2)} - \frac{g^1}{T}dn^1 - \frac{g^2}{T}dn^2 \Rightarrow g^1 = g^2$$

For a three-phase one-component system, each pair of phases should be in equilibrium for the whole system to be in equilibrium; so, the conditions are

$$T^1 = T^2 = T^3 \tag{6.4}$$

$$P^1 = P^2 = P^3 \tag{6.5}$$

$$g^1 = g^2 = g^3 \tag{6.6}$$

6.3.2. Def nition of Heterogeneous States

As a consequence of the equilibrium conditions, P and T in an heterogeneous state does not def ne unambiguously the state, because there are inf nite states with the same P and T; and that differ on the mass fraction of the phases. Nevertheless, two independent variables are enough to def ne such states (but P and T are not independent in heterogeneous systems!); for example, P and a mass fraction.

To show how all the other thermodynamic functions can be obtained from these, we must remember that P (or T) f x the specif c properties of both phases in equilibrium (let's call them z^1 and z^2); for *any extensive property* Z, we can consider the mean specif c property $\mathsf{z} = \mathsf{Z}/m$, where m is the total mass. Any extensive property is additive:

$$\mathsf{Z} = \mathsf{Z}^1 + \mathsf{Z}^2 \Rightarrow \mathsf{z} = \frac{\mathsf{Z}}{m} = \frac{m^1\mathsf{z}^1}{m} + \frac{m^2\mathsf{z}^2}{m} = x\mathsf{z}^1 + (1-x)\mathsf{z}^2$$

where $x = m^1/m$ is the mass fraction of phase 1, and therefore $m^2/m = 1 - x$ is the mass fraction of phase 2.

For example

$$v = xv^1 + (1-x)v^2$$
$$h = xh^1 + (1-x)h^2$$
$$s = xs^1 + (1-x)s^2$$

These relations are valid for any two-phase equilibrium, vapor-liquid, solid-liquid, solid-vapor or solid-solid, with x being in each case the mass fraction of phase 1.

Similarly, for a three-phase one-component system in equilibrium,

$$\mathsf{z} = \frac{\mathsf{Z}}{m} = x^1\mathsf{z}^1 + x^2\mathsf{z}^2 + (1 - x^1 - x^2)\mathsf{z}^3$$
$$v = x^1v^1 + x^2v^2 + (1 - x^1 - x^2)v^3$$
$$h = x^1h^1 + x^2h^2 + (1 - x^1 - x^2)h^3$$
$$s = x^1s^1 + x^2s^2 + (1 - x^1 - x^2)s^3$$

These relations are valid for any three-phase equilibrium that the substance may have: vapor-liquid-solid, solid 1-solid 2-liquid, solid 1-solid 2-solid 3, and so on.

6.3.3. Calculation of Property Changes in a Change of Phase

It is frequently necessary to calculate the difference of a specifc function between the two phases, $\mathsf{z}^1 - \mathsf{z}^2$.

DEFINITION: *Latent heat* is the difference between the specifc enthalpy of one phase and of another in equilibrium.

Latent heats are always taken so as to make them positive (for example, the latent heat of melting is $l_{\text{melt}} = h^L - h^S$, the enthalpy of the liquid less the enthalpy of the solid). We shall consider in the following that phase 1 is the one with higher enthalpy, so $l = h^1 - h^2$. As both phases must be in equilibrium, they are at the same pressure and temperature. The pressure is not independent of temperature in heterogeneous systems, therefore, latent heats can be expressed as a function of only temperature.

The Gibbs function change is, from the equilibrium condition, null: $g^1 - g^2 = 0$. Therefore, the specifc entropy change is:

$$g^1 = h^1 - Ts^1 = g^2 = h^2 - Ts^2 \Rightarrow s^1 - s^2 = \frac{h^1 - h^2}{T} = \frac{l}{T}$$

The specifc volume change can be expressed as

$$v^1 - v^2 = \frac{RT}{P}(z^1 - z^2)$$

which can be obtained as a function of temperature from the thermal equation of state.

Other thermodynamic function changes could be obtained easily from these; for example:

$$u^1 - u^2 = h^1 - h^2 - P(v^1 - v^2) = l - RT(z^1 - z^2)$$
$$a^1 - a^2 = g^1 - g^2 - P(v^1 - v^2) = -RT(z^1 - z^2)$$

The relation between P and T in a two phase system, known as the Clausius-Clapeyron equation, can be obtained from (6.3):

$$\frac{dP}{dT} = \frac{l}{T(v^1 - v^2)} \tag{6.7}$$

where l stands for the *latent heat* of the substance at a given pressure (which implies a unique temperature); ^{1} indicates the phase with higher enthalpy, and ^{2} the other phase.

PROOF

Along the curve of equilibrium in the P-T diagram, $g^1 = g^2$. If we change P by dP, T should change by dT to move to other point where the specif c Gibbs function of both phases are still the same; so, the change of the specif c function of both phases must be the same, $dg^1 = dg^2$. But, from the Gibbs relation,

$$dg^1 = -s^1\,dT + v^1\,dP = dg^2 = -s^2\,dT + v^2\,dP \Rightarrow$$
$$\frac{dP}{dT} = \frac{s^1 - s^2}{v^1 - v^2} = \frac{l}{T(v^1 - v^2)}$$

It is worth noting that this equation can be written as

$$\frac{d\ln P}{d(1/T)} = -\frac{l/R}{z^1 - z^2}$$

that can be integrated making reasonable assumptions about the form of the variation with T of the right-hand side. For example, for not too great variations of T, it might be approximated by a lineal function of T, resulting on

$$\ln P = \mathsf{A} + \frac{\mathsf{B}}{T} + \mathsf{C}\ln T$$

known as Rankine-Kirchoff equation.

6.4. Substances outside Stable Equilibrium

For an isolated system (closed, adiabatic and with no work interchange), entropy cannot decrease. So, if the entropy is at a maximum, with the given conditions (n, U, V constants) there can be no process that lead the system out of this state; the system is at equilibrium,

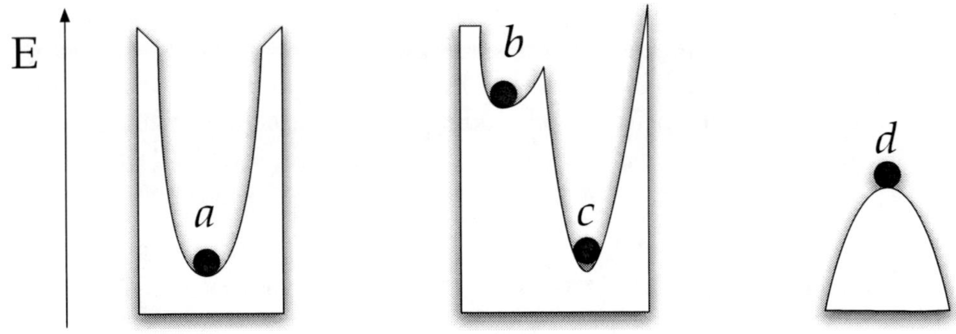

Figure 6.3. Mechanical analogy to illustrate metastable states. E stands for potential energy.

sometimes called stable equilibrium. The maximum entropy condition can be transform in any of the following:

$$U \text{ is minimum at constant } S, V, n \tag{6.8}$$
$$H \text{ is minimum at constant } S, P, n \tag{6.9}$$
$$A \text{ is minimum at constant } T, V, n \tag{6.10}$$
$$G \text{ is minimum at constant } T, P, n \tag{6.11}$$

If the system does not fulf ll these conditions, the initial state cannot last, because the system always will suffer perturbations that will take it out of the initial state. Such equilibrium will be unstable, and it will be impossible to observe it. It can be proved that a system will be unstable if it does not satisfy these requirements:

$$\kappa^{-1} \geq 0 \Rightarrow \kappa \geq 0 \tag{6.12}$$
$$c_v^{-1} \geq 0 \Rightarrow c_v \geq 0 \tag{6.13}$$

In such cases, even the smallest perturbation will take the system def nitively out of its initial state without possible return, as a result of the Second Law. But there can be states that are stable for small perturbations and unstable for bigger ones. Bigger perturbations are less probable, so these states might last long enough to be observed; sometimes, they will be even hard to tell from stable equilibrium, because they might last for really very long time. These states are called *metastable*.

> The f rst condition is the f rst to be broken, so the limit of stability is at $\kappa = \infty$ (spinodal curve). The critical point is a point of limit of stability.

Therefore, a substance in a metastable state $(T,P)_{\text{meta}}$, which has a certain $G_{\text{meta}}(T,P)$, could always rearrange into another state 'seq', which would satisfy $G_{\text{meta}}(T,P) > G_{\text{seq}}(T,P)$. A mechanical analogy can illustrate this point, as in f gure 6.3.

The black dot in the f gure represents a ball. If the ball is conf ned into a valley, as in 'a', by the principle of minimum energy it will tend to maintain its position, even if eventually pushed out of it by external perturbation. Its position at the bottom of the valley is, therefore, a *state of stable equilibrium*. The same would occur if the ball rested on a shallower valley as in point 'b' of the f gure. However, if an external perturbation of suff cient magnitude would occur, the ball could be pushed out of the valley onto a lower one, as indicated by 'c'. The magnitude of the perturbation required for this would be smaller, the shallower the valley. *Unstable equilibrium* is represented in the f gure by 'd'. In this case, even a differential perturbation would push the ball out of its state of equilibrium.

6.5. Example of Thermodynamic Calculation with Heterogeneous System

A rigid closed vessel of $V = 1\,\text{m}^3$ contains $m = 50\,\text{kg}$ of water in vapor-liquid equilibrium at $t_1 = 50\,°\text{C}$. The system is heated until $t_2 = 150\,°\text{C}$.

The following properties of water are known: specif c volume of the saturated liquid at $50\,°\text{C}$: $v_1^L = 1.01 \cdot 10^{-3}\,\text{m}^3/\text{kg}$; at $150\,°\text{C}$: $v_2^L = 1.09 \cdot 10^{-3}\,\text{m}^3/\text{kg}$. For the saturated vapor in the pressures of interest, the specif c volume can be considered the same as in the ideal gas state. Specif c heat at constant pressure for the liquid: $c_p^L = 4.18\,\text{kJ/kg K}$, that can be considered constant in all conditions of interest. For the liquid, the specif c enthalpy and specif c entropy does not change appreciably with pressure at constant temperature in the range of conditions of interest: $\left(\dfrac{\partial h^L}{\partial P}\right)_T \approx 0$, $\left(\dfrac{\partial s^L}{\partial P}\right)_T \approx 0$. The vapor pressure of water is given with suff cient precision by

$$\ln \frac{P}{1\,\text{bar}} = 18.9 - \frac{7046\,\text{K}}{T} - 5.844 \ln \frac{T}{372.78\,\text{K}}$$

Calculate the initial and f nal pressures and vapor qualities, the heat absorbed in the process and the variations of entropy and exergy. Ambient conditions: $T_0 = 300\,\text{K}$, $P_0 = 1\,\text{bar}$

Solution: First, let us note that the assumptions given are reasonable. Pressures in this problem will not be high, the initial been less than 1 bar and the f nal of a few bars, and, as the critical pressure of water is about 220 bar, reduced pressures are quite low, so the compressibility coeff cient of vapor will be close to unity. The vapor pressure given follows the form of the Rankine-Kirchoff equation, that we know that for relatively small ranges of temperatures is precise enough; in this case, the interval is a little bit too big, so results will be a fair approximation, although not precise; an equation of the form of Riedel, for example, would have been better. Finally, entropy in a liquid changes very slowly with pressure, as $\left(\frac{\partial s}{\partial P}\right)_T = -\left(\frac{\partial v}{\partial T}\right)_P$, so for liquid water pressure should change hundreds of bars for the change in entropy to be appreciable. The enthalpy approximation is fair although not as good, because $\left(\frac{\partial h}{\partial P}\right)_T = v(1-\alpha T)$ which is very close to v; so, this approximation is often made in practice, although a change in pressure of 1 bar at constant temperature will imply a change in enthalpy of about 0.1 kJ/kg which, even that small, is exactly the same as the isentropic work needed for this change of pressure.

Care should be taken with units of measure; all specif c units in the given data are per kg, so R must be changed:

$$R = 8.3144\,\frac{\text{J}}{\text{mol K}}\,\frac{1:\text{mol}}{18\cdot 10^{-3}\,\text{kg}} = 461.92\,\text{J/kg K}$$

Substituting initial and f nal temperatures in the vapor pressure equation we f nd the initial and f nal pressures respectively,

$$P_1 = 0.126\,\text{bar} \quad ; \quad P_2 = 4.518\,\text{bar}$$

Vapor quality is determined by the fact that the vessel is neither half-full or half-empty, but always completely full, up to some place of liquid and the rest of vapor, so we have

$$\frac{V}{m} = x_1 v_1^V + (1-x_1)v_1^L \Rightarrow x_1 = \frac{\frac{V}{m} - v_1^L}{\frac{RT_1}{P_1} - v_1^L} = 0.0016$$

$$\frac{V}{m} = x_2 v_2^V + (1-x_2)v_2^L \Rightarrow x_2 = \frac{\frac{V}{m} - v_2^L}{\frac{RT_2}{P_2} - v_2^L} = 0.0438$$

We will need specif c enthalpies and entropies for both phases and states; let us take $h_1^L = 0$ and $s_1^L = 0$ as origin for them. As we can assume that liquid specif c enthalpy and entropy vary only with temperature,

$$h_2^L = c_p^L(T_2 - T_1) = 418\,\text{kJ/kg} \quad ; \quad s_2^L = c_p^L \ln\frac{T_2}{T_1} = 1.1270\,\text{kJ/kg K}$$

Notice that we cannot do something similar with vapor, because we do not know the specif c heat of vapor water. Latent heat can be obtained also from the vapor pressure, remembering the Clausius-Clapeyron equation; differentiating the vapor pressure curve, we get

$$\frac{1}{P}\frac{dP}{dT} = \frac{7046\,\text{K}}{T^2} - \frac{5.844}{T}$$

and with Clausius-Clapeyron and remembering that $v^V \approx RT/P$,

$$\frac{dP}{dT} = \frac{l}{T(v^V - v^L)} \Rightarrow l = \left(\frac{7046\,\text{K}}{T} - 5.844\right)\left(RT - Pv^L\right) \to \begin{cases} l_1 = 2382\,\text{kJ/kg} \\ l_2 = 2107\,\text{kJ/kg} \end{cases}$$

and therefore

$$h_1^V = h_1^L + l_1 = 2382\,\text{kJ/kg} \quad ; \quad s_1^V = s_1^L + \frac{l_1}{T_1} = 7.3712\,\text{kJ/kg K}$$

$$h_2^V = h_2^L + l_2 = 2525\,\text{kJ/kg} \quad ; \quad s_1^V = s_1^L + \frac{l_1}{T_1} = 6.1063\,\text{kJ/kg K}$$

So, f nally

$$Q = \Delta U + \cancelto{0}{W} = \Delta H - V\Delta P$$
$$= m[h_2^L + x_2(h_2^V - h_2^L) - h_1^L - x_1(h_1^V - h_1^L)] - V(P_2 - P_1) = 24.9\,\text{MJ}$$
$$\Delta S = m[s_2^L + x_2(s_2^V - s_2^L) - s_1^L - x_1(s_1^V - s_1^L)] = 66.665\,\text{kJ/K}$$
$$\Delta B = \Delta U - T_0\Delta S + P_0\cancelto{0}{\Delta V} = Q - T_0\Delta s = 4.9\,\text{MJ}$$

Chapter 7

Calculation of Thermodynamic Properties

This chapter is devoted to explaining several practical ways that can be used for calculating properties of real substances: such as h, s, and v for instance. Of course, there always exists the possibility of having a thermal equation and an analytical expression for either c_p or c_v at hand, in which case the engineer has the possibility of deriving the expressions to obtain whichever variable is needed for his balance calculations. However, this happens seldom in practice.

Most of the time, the volume, enthalpy or entropy, of a substance at the required P and T have to be found by relying on reference tables. Nearly always, the necessary value required is not expressed in such a useful way and must be calculated from the referenced reading.

We are going to explain the concept of *discrepancy*, which is as simple as *the difference in the desired property between its actual value and the value it would have if it were an ideal gas at that same* P, T. We shall explain some properties of discrepancies which might be useful for calculations. We shall also explain that there exist discrepancy tables based on generalized thermal equations of substances, and outline how to read them. We shall also explain how to look in the tables of *fugacity*, the quickest way of f nding out the value of g.

7.1. Discrepancies

As we have seen, it is possible to calculate any thermodynamic function from the knowledge of both the thermal equation of state $z(T,P)$ and the isobaric heat capacity at null pressure $c_p^*(T)$. The *method of discrepancies*, based on the ideal gas model, is a very useful tool for the calculation of thermodynamic properties.

7.1.1. Def nitions

DEFINITION: $\mathrm{z}^*(T,P)$ is the value of the thermodynamic function z (any specif c function) that the evaluated system would have at temperature T and pressure P if the system's thermal equation of state were the ideal gas one instead of its true equation of state, maintaining the same c_p^*.

DEFINITION: The discrepancy of function z is def ned as:

$$\mathsf{z}^D = \mathsf{z}(T,P) - \mathsf{z}^*(T,P) \tag{7.1}$$

were z and z* are evaluated *at the same values of T and P.*

It must be remarked that, when a pair of independent variables (x,y) different from (T,P) is chosen for the description, according to the previous def nition z^* must be evaluated at x^* and y^*:

$$\mathsf{z}^D(x,y) = \mathsf{z}(x,y) - \mathsf{z}^*(x^*,y^*) \tag{7.2}$$

Let us state the following proposition. If $\lim_{P\to 0} \mathsf{z}(T,P) = \mathsf{z}^*(T,P)$, then:

$$\mathsf{z}^D = \int_0^P \left[\left(\frac{\partial \mathsf{z}}{\partial P}\right)_T - \left(\frac{\partial \mathsf{z}^*}{\partial P}\right)_T\right] dP \tag{7.3}$$

This expression can be used if a equation of state explicitly expressed in terms of temperature and pressure z(T,P) is known, in order to calculate the derivatives inside the integral.

$$\mathsf{z}^D = \int_\infty^v \left[\left(\frac{\partial \mathsf{z}}{\partial v}\right)_T - \left(\frac{\partial \mathsf{z}^*}{\partial v}\right)_T\right] dv - \int_v^{v^*} \left(\frac{\partial \mathsf{z}^*}{\partial v}\right)_T dv \tag{7.4}$$

This expression can be used if a equation of state explicitly expressed in terms of temperature and specif c volume z(T,v) is known, which would let us obtain the needed derivatives.

PROOF

$$\begin{aligned}\mathsf{z}^D &= \mathsf{z}(T,P) - \mathsf{z}^*(T,P) = [\mathsf{z}(T,P) - \mathsf{z}(T,0)] - [\mathsf{z}^*(T,P) - \mathsf{z}^*(T,0)] \\ &= \int_0^P \left(\frac{\partial \mathsf{z}}{\partial P}\right)_T dP - \int_0^P \left(\frac{\partial \mathsf{z}^*}{\partial P}\right)_T dP = \int_0^P \left[\left(\frac{\partial \mathsf{z}}{\partial P}\right)_T - \left(\frac{\partial \mathsf{z}^*}{\partial P}\right)_T\right] dP\end{aligned}$$

$$\begin{aligned}\mathsf{z}^D &= \mathsf{z}(T,v) - \mathsf{z}^*(T,v^*) = [\mathsf{z}(T,v) - \mathsf{z}^*(T,v)] - [\mathsf{z}^*(T,v^*) - \mathsf{z}^*(T,v)] \\ &= [\mathsf{z}(T,v) - \mathsf{z}(T,\infty)] - [\mathsf{z}^*(T,v) - \mathsf{z}^*(T,\infty)] - [\mathsf{z}^*(T,v^*) - \mathsf{z}^*(T,v)] \\ &= \int_\infty^v \left[\left(\frac{\partial \mathsf{z}}{\partial v}\right)_T - \left(\frac{\partial \mathsf{z}^*}{\partial v}\right)_T\right] dv - \int_v^{v^*} \left(\frac{\partial \mathsf{z}^*}{\partial v}\right)_T dv\end{aligned}$$

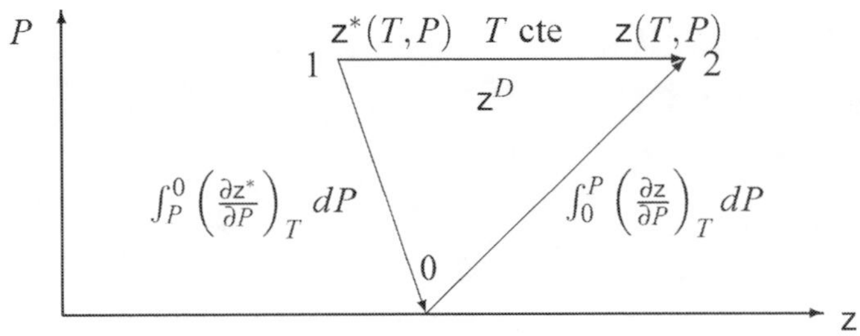

Figure 7.1. Calculation of z-discrepancy as a function of T and P

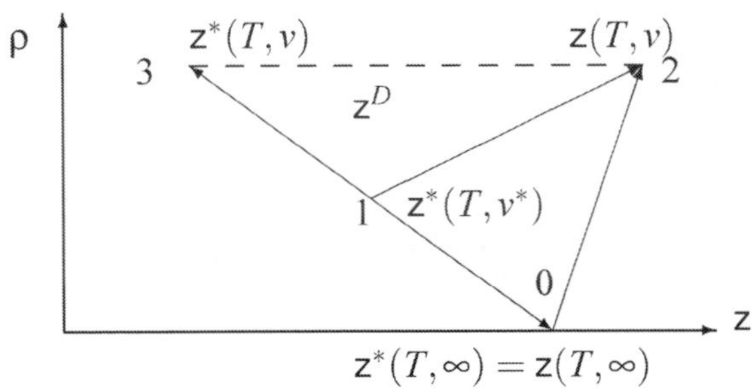

Figure 7.2. Calculation of z-discrepancy as a function of T and v

A graphical presentation of the previous results is given in figures 7.1 and 7.2. In the first one, $\int_P^0 \left(\frac{\partial z^*}{\partial P}\right)_T dP$ and $\int_0^P \left(\frac{\partial z}{\partial P}\right)_T dP$ can be identified with the projections of segments $\overline{10}$ and $\overline{02}$ respectively onto the abscissa axis, and the discrepancy $z^D(T,P)$ would then be the segment $\overline{12}$. In figure 7.2, $\int_v^\infty \left(\frac{\partial z^*}{\partial v}\right)_T dv$, $\int_v^{v^*} \left(\frac{\partial z^*}{\partial v}\right)_T dv$, $\int_\infty^v \left(\frac{\partial z}{\partial v}\right)_T dv$ and $z^D(T,v)$ can be identified with the projections of $\overline{30}$, $\overline{31}$, $\overline{02}$, and $\overline{12}$, respectively.

The hypothesis on which the previous proposition is based is not satisfied by all the intensive functions we can find, as the ideal gas is just a model that behaves like reality only in some aspects. This is the case, among others, of the Joule-Thompson coefficient μ_{JH}, the caloric coefficient λ_T, and the specific volume v (and this is not contradictory with **BL1**!):
A expansion in powers of P of the compressibility factor $z = 1 + \mathrm{b}(T)P + \mathrm{c}(T)P^2 + \dots$ leads to $\lim_{P\to 0}(v - v^*) = \lim_{P\to 0}\left[\frac{RT}{P}(z-1)\right] = RT\mathrm{b}(T) \neq 0$

7.1.2. Illustrative Examples

Example 1 Let us calculate h^D and g^D for a gas with a virial equation of state, expressed in terms of powers of P, with three terms $z = 1 + \mathsf{b}(T)P + \mathsf{c}(T)P^2$.

$$\left(\frac{\partial h}{\partial P}\right)_T = v - T\left(\frac{\partial v}{\partial T}\right)_P \quad ; \quad \left(\frac{\partial h^*}{\partial P}\right)_T = 0$$

$$v = \frac{RT}{P} + RT\mathsf{b}(T) + RT\mathsf{c}(T)P$$

$$\left(\frac{\partial v}{\partial T}\right)_P = \frac{R}{P} + R\mathsf{b}(T) + RT\mathsf{b}'(T) + R\mathsf{c}(T)P + RT\mathsf{c}'(T)P$$

$$\left(\frac{\partial h}{\partial P}\right)_T = -RT^2\left[\mathsf{b}'(T) + \mathsf{c}'(T)P\right]$$

$$h^D = \int_0^P \left[\left(\frac{\partial h}{\partial P}\right)_T - \left(\frac{\partial h^*}{\partial P}\right)_T\right] dP = -RT^2\left[\mathsf{b}'(T)P + \frac{1}{2}\mathsf{c}'(T)P^2\right]$$

$$\left(\frac{\partial g}{\partial P}\right)_T = v = \frac{RT}{P}z \quad ; \quad \left(\frac{\partial g^*}{\partial P}\right)_T = v^* = \frac{RT}{P}$$

$$g^D = \int_0^P \left[\left(\frac{\partial g}{\partial P}\right)_T - \left(\frac{\partial g^*}{\partial P}\right)_T\right] dP = RT\int_0^P \frac{z-1}{P} dP = RT\left[\mathsf{b}(T)P + \frac{1}{2}\mathsf{c}(T)P^2\right]$$

Example 2 Let us calculate s^D for a gas with van der Waals equation of state $(P + \frac{\mathsf{a}}{v^2})(v - \mathsf{b}) = RT$.

In this case, it is not easy to express v explicitly in terms of T and P. In spite of this, we can choose the description in terms of the independent variables T and v:

$$P(T,v) = \frac{RT}{v - \mathsf{b}} - \frac{\mathsf{a}}{v^2}$$

$$\left(\frac{\partial s}{\partial v}\right)_T = \left(\frac{\partial P}{\partial T}\right)_v = \frac{R}{v - \mathsf{b}} \quad ; \quad \left(\frac{\partial s^*}{\partial v}\right)_T = \frac{R}{v}$$

$$\begin{aligned} s^D(T,v) &= \int_\infty^v \left[\left(\frac{\partial s}{\partial v}\right)_T - \left(\frac{\partial s^*}{\partial v}\right)_T\right] dv - \int_v^{v^*} \left(\frac{\partial s^*}{\partial v}\right)_T dv \\ &= \int_\infty^v \left(\frac{R}{v-\mathsf{b}} - \frac{R}{v}\right) dv - \int_v^{v^*} \frac{R}{v} dv = R\ln\frac{v-\mathsf{b}}{v} - R\ln\frac{v^*}{v} \\ &= R\ln\frac{v-\mathsf{b}}{v^*} = \ln\frac{v-\mathsf{b}}{\frac{RT}{P}} = \ln\left[1 - \frac{\mathsf{a}(v-\mathsf{b})}{RTv^2}\right] \end{aligned}$$

7.2. Fugacity

DEFINITION: Fugacity f is the function of state defined by the following identity:

$$g(T,P) = g^*(T,P^u) + RT\ln\frac{f(T,P)}{P^u} \tag{7.5}$$

where P^u is an arbitrary pressure (frequently 1 bar, 1 Pa, 1 ata, 1 mmHg...), whose value is completely indifferent.

DEFINITION: The fugacity coeff cient is the dimensionless number:

$$\phi(T,P) = \frac{f(T,P)}{P} \tag{7.6}$$

Now we shall show that the previous def nition of fugacity is independent of the arbitrary pressure P^u chosen:

PROOF

$$g(T,P) = g^*(T,P_u) + RT\ln\frac{f(T,P)}{P_u}$$
$$g(T,P) = g^*(T,P_u') + RT\ln\frac{f'(T,P)}{P_u'}$$

Subtracting:

$$RT\ln\frac{f}{f'} - RT\ln\frac{P_u}{P_u'} + g^*(T,P_u) - g^*(T,P_u') = 0$$

and taking into account that:

$$g^*(T,P_u) - g^*(T,P_u') = \int_{P_u'}^{P_u}\left(\frac{\partial g^*}{\partial P}\right)_T dP = \int_{P_u'}^{P_u} v^* dP = \int_{P_u'}^{P_u}\frac{RT}{P}dP = RT\ln\frac{P_u}{P_u'}$$

the equality between f and f' has been proved.

A relation between the fugacity coeff cient and the discrepancy of the specif c Gibbs function arises:

$$\ln\phi = \frac{g^D}{RT} \tag{7.7}$$

PROOF

$$g(T,P) = g^*(T,P_u) + RT\ln\frac{f}{P_u}$$
$$g^*(T,P) = g^*(T,P_u) + RT\ln\frac{P}{P_u}$$

Subtracting:

$$RT\ln\frac{f}{P} = g(T,P) - g^*(T,P) = g^D$$

Relation (7.7) allows us to calculate fugacity from the method given in the preceding exposition (see Example 1 of section 7.1.2.). The virial equation, expressed as $z = 1 + \mathsf{b}(T)P + \mathsf{c}(T)P^2 + \ldots$, allows us to get:

$$\ln\phi = \mathsf{b}(T)P + \frac{1}{2}\mathsf{c}(T)P^2 + \ldots$$

$$f = P\exp\left[\mathsf{b}(T)P + \frac{1}{2}\mathsf{c}(T)P^2 + \ldots\right]$$

7.2.1. Variation of Fugacity

$$d\ln f = -\frac{h^D}{RT^2}dT + \frac{v}{RT}dP \qquad (7.8)$$

$$d\ln\phi = -\frac{h^D}{RT^2}dT + \frac{v^D}{RT}dP \qquad (7.9)$$

PROOF

Deriving (7.5) respect to P, $v = RT\left(\frac{\partial \ln f}{\partial P}\right)_T$ is given, so then:

$$\left(\frac{\partial \ln f}{\partial P}\right)_T = \frac{v}{RT}$$

Dividing (7.5) by T, deriving it respect to T, applying the Gibbs-Helmholtz relation $\left(\frac{\partial(g/T)}{\partial T}\right)_P = -\frac{h}{T^2}$ and considering that h^* is independent of P it is now obtained $-\frac{h}{T^2} = -\frac{h^*}{T^2} + R\left(\frac{\partial \ln f}{\partial T}\right)_P$, leading to:

$$\left(\frac{\partial \ln f}{\partial T}\right)_P = -\frac{h^D}{RT^2}$$

$$\left(\frac{\partial \ln\phi}{\partial P}\right)_T = \left(\frac{\partial \ln f}{\partial P}\right)_T - \frac{1}{P} = \frac{v}{RT} - \frac{v^*}{RT} = \frac{v^D}{RT}$$

$$\left(\frac{\partial \ln\phi}{\partial T}\right)_P = \left(\frac{\partial \ln f}{\partial T}\right)_P = -\frac{h^D}{RT^2}$$

7.2.2. Equilibrium between Phases

In section 6.3.1. of chapter 6 we saw that, for a system with only one chemical component, the general condition for thermodynamic equilibrium between two phases 1 and 2 is:

$$P^1 = P^2 \quad ; \quad T^1 = T^2 \quad ; \quad g^1 = g^2$$

According to the fugacity def nition (7.5), the last condition can be substituted by:

$$f^1 = f^2 \quad \text{or equivalently} \quad \phi^1 = \phi^2 \tag{7.10}$$

This condition allows to analytically obtain the equilibrium condition in a phase transition $P^s = P(T)$ from an equation of state similar to the kind found in van der Waals' equation; an alternative way to Clausius-Clapeyron equation.

7.2.3. Fugacity of Condensed Phases

To obtain fugacity of condensed phases (solid or liquid), the calculation is simplif ed keeping in mind that, for them, the specif c volume is almost independent of pressure. For the case of a liquid, if the vapor pressure curve is known, we have:

$$f^{L,\text{sat}} = f^{V,\text{sat}} = \phi^{V,\text{sat}} P^{\text{sat}}$$

Integrating (7.8) we get:

$$\ln \frac{f^L(T,P)}{f^L(T,P^{\text{sat}})} = \int_{P^{\text{sat}}}^{P} \frac{v^L}{RT} dP \approx \frac{v^L}{RT}(P - P^{\text{sat}})$$

$$f^L(T,P) \approx \phi^{V,\text{sat}} P^{\text{sat}}(T) \exp\left[\frac{v^L}{RT}(P - P^{\text{sat}})\right] \tag{7.11}$$

We can group this in the following way:

$$f^L(T,P) \approx P^{\text{sat}}(T)\Pi(T,P) \tag{7.12}$$

were the factor $\Pi(T,P) = \phi^{V,\text{sat}} \exp\left[\frac{v^L}{RT}(P - P^{\text{sat}})\right]$ is called *Pointing correction*, and it is quite close to unity for moderate equilibrium pressures ($\phi^V \approx 1$; $v^L \ll \frac{RT}{P}$):

$$f^L(T,P) \approx P^{\text{sat}}(T) \quad \text{(moderate pressure)} \tag{7.13}$$

7.3. Calculation of Thermodynamic Properties from Generalized Equations

7.3.1. Tabular Data for the Compressibility Factor

If the *principle of corresponding states* were valid, a universal relation to express the compressibility factor as a function of only two parameters would exist, $z = z(T_r, P_r)$, where $T_r = \frac{T}{T_c}$, and $P_r = \frac{P}{P_c}$ are the reduced pressure and temperature, and the two parameters are the critic properties T_c and P_c. However, that principle is only satisf ed in an approximate way, and hence it is necessary to introduce a third parameter in order to achieve a suff ciently precise description of pure substances' equations of state. In practice, the following factor is used.

Critic Compressibility Factor

Experience shows that z_c varies between 0.22 and 0.29 for the different substances, with a modal value of 0.27. Lydersen-Greenkorn-Hougen[1] prepared tables for $z_c = 0.27$ and correcting factors for $z_c > 0.27$ and $z_c < 0.27$, getting a good adjustment to the experimental data for both, vapor and liquid. Those tables have a double input: in the columns' heading the reduced pressure values are listed and in the rows' one the reduced temperature is set. The values of z that appear in the table are the corresponding ones to $z_c = 0.27$. For a different z_c, a correction must be applied:

$$z = z_{\text{read from table}} + \mathsf{D}(z_c - 0.27) \tag{7.14}$$

Together with z, two additional sets of data are given: D_a, which should be used for the previous correction when $z_c > 0.27$, and D_b, and should be taken when $z_c < 0.27$. D values can be found in the tables up to a reduced pressure of 1.2. For higher pressures, and for states with $T_r < 0.8$ or with $T_r > 1.2$, the $P_r = 1.2$ values for D can be used. In the rest of the cases, a interpolation is not recommended. For quantum gases, H_2, He and Ne, modifed critic constants have been proposed, since the true ones do not produce good predictions.

Just below the reduced pressure, the values of T_r^s (reduced saturation temperature), z^{Vs} (compressibility factor of saturated vapor at that pressure) and z^{Ls} (compressibility factor of saturated liquid at that pressure), are given for each column. In the columns with $P_r < 1$ the liquid zone and the vapor zone are separated by a horizontal line. One of the drawbacks of these tables is that they do not provide a method to correct the saturation curve for the cases with $z_c \neq 0.27$.

Generally, when the value of interest for T_r or P_r is not included in the tables, it can be applied to a linear interpolation, but it must be remarked that only a interpolation between states in the same state of aggregation is correct.

Acentric Factor

Pitzer noticed that many gases with spheric molecular symmetry present a reduced saturation pressure of 0.1 at reduced temperature of 0.7. He then introduced the acentric factor, defned as $\omega = -1 - \log P_r^s|_{T_r=0.7}$. So, then ω becomes null for a gas constituted by spherical molecules. The different values of ω have been tabulated for a large number of substances.

The calculation of the compressibility factor z is carried out by means of[2]

$$z = z^{(0)}(P_r, T_r) + \omega \cdot z^{(1)}(P_r, T_r) \tag{7.15}$$

where $z^{(0)}$ and $z^{(1)}$ are found in tables. The original Pitzer & Curl's tables were extended later by Lee and Kesler, increasing its range of application, making it signifcantly larger than the one given in Lydersen's tables. Lee-Kesler tables are equivalent to the analytic Lee-Kesler thermal equation of state. They are double-input tables, with reduced pressure values in the column's headings and reduced temperature values in the row's headings, in a

[1]A. L. Lydersen, R. A. Greenkorn, O. A. Hougen: 'Generalized Thermodynamic Properties of Pure Fluids', Univ. Wisconsin Coll. Eng., Eng. Exp. Stn. Rept. 4 (1955).

[2]K. S. Pitzer, D. Z. Lippmann, R. F. Curl, C. M. Huggins, D. E. Peterson: 'The Volumetric and Thermodynamic Properties of Fluids II', J. Am. Chem. Soc. 77, pp. 3433-3440 (1955).

similar way to the Lydersen's tables. For the calculation of saturation pressure or saturation temperature, the Lee-Kesler equation for saturation curve must be used. To calculate z for saturated vapors or liquids, a extrapolation (or interpolation in some exceptional cases) from the values read from tables for that saturation pressure or temperature should be carried out. The liquid and vapor zones are separated by a horizontal line.

7.3.2. Tabular Data for Discrepancies and Fugacity Coeff cient

For the calculation of discrepancies from tabular thermal equations of state, it could be followed, in principle, the same method used for the analytical equations of state. But, for this case, the integrals an derivatives needed could result in a tedious work, as we would need to use numerical methods for obtaining them. For that reason, also some discrepancy values have been tabulated directly, for Lydersen tabular equation of state (using z_c as the third parameter) and for Lee-Kesler equation (using ω as the third parameter) in a similar way to the compressibility factor tables.

In the Lydersen tables (z_c), $-\frac{h^D}{T_c}$ and $-s^D$ have been tabulated, both in cal/(mol·K). Corrections for the case with $z_c \neq 0.27$ are done in a similar way to the compressibility factor tables:

$$-\frac{h^D}{T_c} = \left(-\frac{h^D}{T_c}\right)_{\text{read from table}} + \mathsf{D}'(z_c - 0.27) \tag{7.16}$$

$$-s^D = (-s^D)_{\text{read from table}} + \mathsf{D}''(z_c - 0.27) \tag{7.17}$$

where D' and D'' are tabulated for every case, as well. Additionally, for $P_r < 1$, the reduced saturation temperature and the values of the tabulated function for the saturated liquid and vapor are given in the table (only valid for the reference case $z_c = 0.27$). Lydersen tables provide data for the fugacity coeff cient ϕ as well (even it is not strictly needed since ϕ could be related to g^D, and this last one could be calculated from the tables for h^D and s^D). In this case the correction procedure is given by:

$$\phi = \phi_{\text{r.f.t}} 10^{\mathsf{D}'''(z_c - 0.27)} \tag{7.18}$$

In the Lee-Kesler tables (ω) the dimensionless values of $\frac{-h^D}{RT_c}$ and $\frac{-s^D}{R}$ can be found. In a similar way to the case of the compressibility factor tables, there are two different tables for each function, characterized by the super-indexes $^{(0)}$ and $^{(1)}$, and the discrepancies are calculated as follows:

$$-\frac{h^D}{RT_c} = \left(-\frac{h^D}{RT_c}\right)^{(0)} + \omega\left(-\frac{h^D}{RT_c}\right)^{(1)} \tag{7.19}$$

$$-\frac{s^D}{R} = \left(-\frac{s^D}{R}\right)^{(0)} + \omega\left(-\frac{s^D}{R}\right)^{(1)} \tag{7.20}$$

For calculating the saturation values of the tabulated functions, the same extrapolation method mentioned for the compressibility factor table must be followed. Finally, also tables for $\log_{10}\phi$ are given and the fugacity coeff cient is obtained by:

$$\phi = 10^{(\log_{10}\phi)^{(0)} + \omega(\log_{10}\phi)^{(1)}} \tag{7.21}$$

Exercise Calculate the specif c work needed to compress CO_2 along a isothermal reversible process going from 8 bar to 55 bar at temperature 298 K.

Data: $T_c = 304.1$ K, $P_c = 73.8$ bar, $z_c = 0.274$, $\omega = 0.239$.

Solution Combining First and Second Laws:

$$\left.\begin{array}{l}\Delta u = q - w \\ \Delta s = \frac{q}{T}\end{array}\right\} \Rightarrow w = -\Delta u + T\Delta s = -\Delta h + \Delta(Pv) + T\Delta s$$

$$\Delta h = \cancel{\Delta h^*} + \Delta h^D = h_2^D - h_1^D$$

$$\Delta(Pv) = \Delta(RTz) = RT(z_2 - z_1)$$

$$\Delta s = \Delta s^* + \Delta s^D = -R\ln\frac{P_2}{P_1} + s_2^D - s_1^D$$

Thus, the whole problem comes down to calculating z_1, z_2, h_1^D, h_2^D, s_1^D and s_2^D from the tabulated data. For the states 1 and 2, we have:

$$T_r = \frac{T}{T_c} = 0.98 \quad ; \quad P_{r1} = \frac{P_1}{P_c} = 0.10840 \quad ; \quad P_{r2} = \frac{P_2}{P_c} = 0.74526$$

Lydersen tables. The following values are read from tables:

P_r	z	D_a	$-\frac{h^D}{T_c}$ (cal/(mol·K))	D'_a	$-s^D$ (cal/(mol·K))	D''_a
0.1	0.961	0.16	0.24	−2.0	0.21	−2.1
0.2	0.921	0.28	0.51	−5.0	0.4	−4.0
0.7	0.665	0.68	2.30	−17.0	1.81	−15.0
0.8	0.580	0.76	2.98	−16.0	2.40	−19.0

Introducing the corrections given by (7.14), (7.16) and (7.17)

P_r	z	$-\frac{h^D}{T_c}$ (cal/(mol·K))	$-s^D$ (cal/(mol·K))
0.1	0.96164	0.232	0.202
0.2	0.92212	0.490	0.384
0.7	0.66772	2.232	1.750
0.8	0.58304	2.916	2.324

After interpolating

P_r	z	$-\frac{h^D}{T_c}$ (cal/mol·K)	$-s^D$ (cal/mol·K)
0.10840	0.95832	0.25367	0.21692
0.74526	0.62940	2.5426	2.0098

Lee-Kesler tables. We read:

P_r	$z^{(0)}$	$z^{(1)}$	$\left(-\frac{h^D}{RT_c}\right)^{(0)}$	$\left(-\frac{h^D}{RT_c}\right)^{(1)}$	$\left(-\frac{s^D}{R}\right)^{(0)}$	$\left(-\frac{s^D}{R}\right)^{(1)}$
0.1	0.9637	−0.0090	0.107	0.101	0.073	0.094
0.2	0.9253	−0.0184	0.221	0.209	0.153	0.196
0.6	0.7360	−0.0641	0.797	0.776	0.580	0.734
0.8	0.5887	−0.1100	1.273	1.324	0.971	1.270

Using (7.15), (7.19) and (7.20)

P_r	z	$-\frac{h^D}{RT_c}$	$-\frac{s^D}{R}$
0.1	0.961 55	0.131 14	0.095 47
0.2	0.920 90	0.270 95	0.199 84
0.6	0.720 68	0.982 46	0.755 43
0.8	0.562 41	1.589 4	1.274 5

and interpolating

P_r	z	$-\frac{h^D}{RT_c}$	$-\frac{s^D}{R}$
0.108 40	0.958 13	0.142 88	0.104 23
0.745 26	0.605 73	1.423 3	1.132 4

Finally,

$$w = -(h_2^D - h_1^D) + RT(z_2 - z_1) - RT\ln\frac{P_2}{P_1} + T(s_2^D - s_1^D) = \begin{cases} -4916.1\,\text{J/mol}, & \text{for a)} \\ -4960.1\,\text{J/mol}, & \text{for b)} \end{cases}$$

From IUPAC tables, the result $w = -4945.6\,\text{J/mol}$ is obtained, so the generalized equations given by Lydersen and Lee-Kesler tables provide a prediction with a high level of accuracy (relative errors of a little less than 0.6% and 0.3% respectively in this example).

Chapter 8

Open Systems

8.1. Introduction

8.1.1. General Considerations

Let us remind that an open system is the kind of system that exchanges mass with its surroundings. The majority of technical processes take part in open systems, that's why it is essential to study them.

> DEFINITION: It is said that the *principle of local equilibrium* is satisf ed when temperature and pressure are def ned at every point within the system, with a continuous variation in space.

Thus, the local equilibrium principle is not a principle at all, but just a condition that sometimes is fulf lled.

In most industrial applications, the considered open system is not in an internal equilibrium state, since it is not at a unique temperature nor pressure across the whole system. However, often the principle of local equilibrium holds. If that is the case, the concepts used here can be extended, just by considering the system as a compound system formed by an inf nite number of elemental subsystems. The intensive thermodynamic variables would then be def ned at every point, and, regarding the additivity property, the value of the extensive variables could be calculated by addition, that is by the integral along the inf nite elemental subsystems.

> DEFINITION: A process in an open system is said to be *quasistatic* when the local equilibrium principle holds for the system at any time.

Although most of the processes with practical interest in industrial facilities are quasistatic processes, there are some non-static cases, such as for example lamination. The general equations for an open system that are going to be stated will be also be valid with the only condition that the principle of local equilibrium is satisf ed in the system's inlet and outlet sections and hence the thermodynamic variables are def ned for those sections.

DEFINITION: A process is said to be *stationary* when the state variables of the open system are time independent.

That is the case of any continuous process in a steady regime. The time-independence condition, is not fully satisfied in practice; however, if the variations are cyclic, the process may be considered stationary in a time interval big enough in relation to the cycle period (e.g., alternative compressors with pressure regulators). When the open system's properties vary in time, the processes is *non-stationary*. There are several practical cases of this kind of processes: filling and discharging of vessels, transient regime in continuous flows, etc.

8.1.2. General Methodology

We shall solve the general case of an open system with a number of mass flow inlets and outlets, by transforming it into a closed system composed of the open system under study plus the differentials of mass that enter and/or exit the system in a differential of time dt (see figure 8.1). The limit of the closed system is denoted by double lines where there are no physical walls.

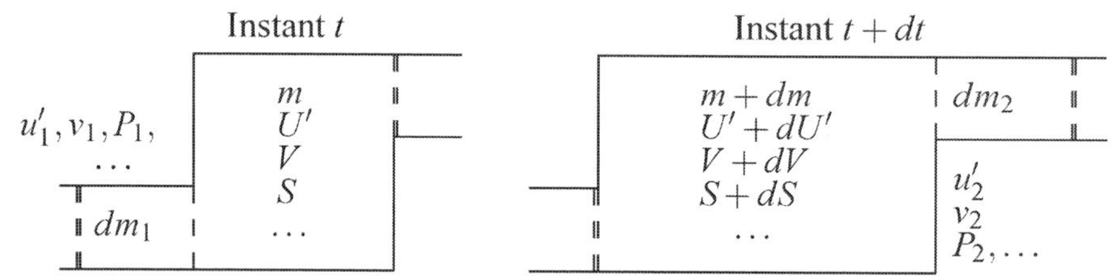

Figure 8.1.

The **open system's variables** will be denoted **with no subindex**, the mass flow inlets variables use subindex 1, and the mass flow outlets are referred to by subindex 2. For that closed system, we can apply the known thermodynamic equations in order to establish the balance for an open system.

8.2. Mass Balance

$$\boxed{dm = \sum_j dm_j} \tag{8.1}$$

Where subindex j represents both, entering or exiting mass flows, with the criterion of considering an entering mass flow equal to its absolute value (positive dm) and an exiting mass flow opposite to its absolute value (negative dm). Equivalently, dividing by dt and using the notation $\dot{Z} = dZ/dt$,

$$\dot{m} = \sum_j \dot{m}_j$$

which simply expresses that *the mass rate of variation of the open system is the algebraic sum of its mass exchanges with the surroundings.*

PROOF

Since we do not take into account nuclear reactions and the relativistic conversion between mass and energy, the closed system's mass remains constant (by definition of closed system).

Mass at instant t:	$m+dm_1$
Mass at instant $t+dt$:	$m+dm+dm_2$
Mass balance:	$m+dm_1 = m+dm+dm_2 \Rightarrow dm = dm_1 - dm_2$

Considering the previously mentioned sign criterion for entering and exiting mass flows, and generalizing to any number of inlets and outlets, we get expression (8.1).

8.3. Energy Balance

In open systems, work due to a volume variation may exist ($dW = P_e dV$), in a similar way to a closed system. However, the axial work exchange, generally referred to as *technical work* and which is not related to a volume variation, is the most usual. Obviously, the technical work is useful work entirely, contrary to what occurs with the volume variation work.

The First Law for open systems reads

$$\boxed{dU' = dQ - dW + \sum_j h'_j dm_j} \tag{8.2}$$

$$\dot{U}' = \dot{Q} - \dot{W} + \sum_j h'_j \dot{m}_j$$

PROOF

Referring to figure 8.1, and calling dQ' the heat exchanged by the closed system in the differential time interval dt and dW the work exchanged by the open system in the same time interval (as axial work or as work due to a volume variation, or any other work exchange), we have:

- the work exchanged by the closed system is (since it is an additive magnitude) the sum of the work exchanged by the open system dW plus the work exerted by the flows on the entering and exiting differentials of mass: $dW + P_2 v_2 dm_2 - P_1 v_1 dm_1$,
- the total energy at instant t is $U' + u'_1 dm_1$,
- the total energy at instant $t+dt$ is $U' + dU' + u'_2 dm_2$.

Hence, applying the general form of the First Law to the closed system,

$$U' + dU' + u'_2 dm_2 - (U' + u'_1 dm_1) = dQ' - dW - P_2 v_2 dm_2 + P_1 v_1 dm_1$$
$$dU' = dQ' - dW + (u'_1 + P_1 v_1) dm_1 - (u'_2 + P_2 v_2) dm_2$$

Taking into account $h = u + Pv$, introducing $h' = h + e_c + e_p = u' + Pv$ and generalizing to any number of inlets and outlets, in equal form to what we did before, we have

$$dU' = dQ' - dW + \sum_j h'_j dm_j$$

Now we call dQ the heat exchange through the open system's external housing. dQ' includes, additionally, the heat exchanged by the entering and exiting differentials of mass with the surroundings. The difference between them is:

- the heat exchanged by the differentials of mass a through the differential of 'housing' element. It would then be an infnitesimal of higher order than dQ.
- the heat exchanged through the ideal surfaces that close the system (inlet and outlet sections). This contribution is negligible in most of the applications for two reasons:
  - the thermal gradients and heat transmission within the fuid are small,
  - the mentioned ideal surfaces use to be small compared to the open system boundary.

Thus, in general $Q' = Q$ and we get (8.2).

8.4. Entropy Balance

The Second Law for open systems:

$$\boxed{dS = dJ_s + d\sigma + \sum_j s_j dm_j} \tag{8.3}$$

$$\dot{S} = \dot{J}_s + \dot{\sigma} + \sum_j s_j \dot{m}_j$$

PROOF

Closed system's entropy at instant t: $S + s_1 dm_1$

Closed system's entropy at instant $t + dt$: $S + dS + s_2 dm_2$

$$S + dS + s_2 dm_2 - S - s_1 dm_1 = dJ_s + d\sigma$$

neglecting higher order infnitesimals. So then

$$dS = dJ_s + d\sigma + s_1 dm_1 - s_2 dm_2$$

and generalizing to any number of inlets and outlets, (8.3) is obtained.

If the open system does not have a unique temperature, but the temperature changes continuously from a point to another one across the system, the entropy f ux on the open system would be the integral of the contributions of every differential element of its boundary in the process (exactly as it would be done for a compound system). Therefore, it would not be $dJ_s = dQ/T$ since there is not an only T, but it would be a surface integral of the second order differentials (in time and in the surface) of the exchanged heat divided by the local temperature. In the case of a non-static process, for which temperature may not even be locally def ned, entropy f ux cannot be separated from generated entropy (except for the particular case of an adiabatic process, where it can be said that the entropy change identif es with the generated entropy). In any case, the Clausius inequality remains valid.

8.5. Flow Exergy

DEFINITION: The so-called *f ow exergy* e and *total f ow exergy* e' are def ned as

$$e \equiv (h - h_0) - T_0(s - s_0) = h - T_0 s - g_0 \tag{8.4}$$

$$e' \equiv e + e_c + e_p = h' - T_0 s - g_0 \tag{8.5}$$

8.5.1. Exergy Balance

$$\boxed{dB' = (dQ - T_0 dJ_s) - (dW - P_0 dV) - T_0 d\sigma + \sum_j e'_j dm_j} \tag{8.6}$$

PROOF

Combining equations (8.2) and (8.3)

$$dU' = dQ - dW + \sum h'_j dm_j \quad ; \quad dS = dJ_s + d\sigma + \sum s_j dm_j$$

we get $$dU' - T_0 dS = (dQ - T_0 dJ_s) - dW - T_0 d\sigma + \sum_j (h'_j - T_0 s_j) dm_j$$

On the other hand, taking into account that $dG_0 = g_0 dm$ and making use of (8.1),

$$dB' = d(U' - T_0 S + P_0 V - G_0) = d(U' - T_0 S) + P_0 dV - g_0 \sum_j dm_j$$

$$= (dQ - T_0 dJ_s) - (dW - P_0 dV) - T_0 d\sigma + \sum_j (h'_j - T_0 s_j - g_0) dm_j$$

which, together with the f ux exergy def nition leads to balance (8.6).

It must be observed that, similarly to the exergy balance of a closed system, the physical meaning of the terms in equation (8.6), is:

- $dQ - T_0 dJ_s$: heat's exergy content (equal to $dQ\left(1-\frac{T_0}{T}\right)$ when the open system is at a unique temperature T).
- $dW - P_0 dV$: work's exergy content.
- $T_0 d\sigma = dI$: Exergy destruction.

8.5.2. Physical Meaning of Flow Exergy

Flow exergy quantif es the total technical work (useful work), per unit mass, that can be obtained from a f ow stream in an open system that follows a stationary and reversible process with a thermal interaction only with the environment, and recovering a f ow stream in the inert state and with no kinetic and potential energy.

PROOF

Applying the exergy balance (8.6) to an open system following a stationary process with a thermal interaction only with the environment, and with only one entering and one exiting stream, where the latter is found in the inert conditions (T_0 and P_0) and with no kinetic and potential energy, we have

$$dB' = 0 = dQ - T_0 dJ_s - dW - dI + (e'_1 - e'_0)dm_1 = dQ - T_0 dJ_s - dW - T_0 d\sigma + e'_1 dm_1$$

since e'_0 is null. On the other hand, if J_{s_0} is the entropy f ux seen by the environment,

$$T_0 dJ_{s_0} = -dQ \quad ; \quad dJ_s + dJ_{s_0} = d\sigma^e \Rightarrow T_0 dJ_s = T_0 d\sigma^e - T_0 dJ_{s_0} = T_0 d\sigma^e + dQ$$
$$\Rightarrow dW = dQ - T_0 dJ_s - T_0 d\sigma + e'_1 dm_1 = e'_1 dm_1 - T_0(d\sigma^e + d\sigma)$$

As $\sigma + \sigma^e \geq 0$, the maximum possible work is obtained for a globally reversible process

$$e'_1 dm_1 = dW^R \Longrightarrow e'_1 = \frac{dW^R}{dm_1} \tag{8.7}$$

Notice that, contrary to the case of exergy B, f ow exergy e could be negative in states with $P < P_0$.

8.6. General Equations in Stationary Processes

Let us keep in mind that in a stationary process the open system's state variables do not vary in time. In that case, in particular the system's mass and energy remain constant; mass f ow, temperature and pressure at inlets and outlets are constant as well; and power and heat exchanged per unit time are constant.

General equations (8.1), (8.2), (8.3) and (8.6) become:

$$dm = 0 \Rightarrow \boxed{0 = \sum_j dm_j} \tag{8.8}$$

$$dU' = 0 \Rightarrow \boxed{0 = dQ - dW + \sum_j h'_j dm_j} \tag{8.9}$$

$$dS = 0 \Rightarrow \boxed{0 = dJ_s + d\sigma + \sum_j s_j dm_j} \tag{8.10}$$

$$dB' = 0; dV = 0 \Rightarrow \boxed{0 = dQ - T_0 dJ_s - dW - dI + \sum_j e'_j dm_j} \tag{8.11}$$

It must be remarked that, in the previous expressions, the work cannot be of volume variation but technical work, since $dV = 0$. Regarding the term dJ_s, in the case that the open system is not in internal equilibrium, it is pertinent what said in the previous section.

For the very common case of only one inlet and one outlet, equation (8.8) just express $dm_1 = -dm_2$, and introducing $\frac{dQ}{dm_1} = q$, $\frac{dW}{dm_1} = w$, $\frac{d\sigma}{dm_1} = \hat{\sigma}$, $\frac{dJ_s}{dm_1} = j_s$, the other balances read

$$h'_2 - h'_1 = q - w \tag{8.12}$$

$$s_2 - s_1 = j_s + \hat{\sigma} \tag{8.13}$$

$$e'_2 - e'_1 = q - T_0 j_s - w - T_0 \hat{\sigma} \tag{8.14}$$

The physical meaning of the terms in (8.14) is:

$q - T_0 j_s = \int_1^2 dq \left(1 - \frac{T_0}{T}\right)$: heat's exergy content.

w : work, equal to work's exergy content.

$T_0 \hat{\sigma} = i$: exergy destruction.

8.7. Momentum Balance

The practical forms of this equation, for a system with one input stream 1 and one output stream 2, are

$$w = -\int_1^2 v dP - \Delta(e_c + e_p) - \int_1^2 T d\hat{\sigma} \tag{8.15}$$

$$\boxed{w^R = -\int_1^2 v dP - \Delta(e_c + e_p)} \tag{8.16}$$

where superindex ^{R} means 'for a reversible process'.

PROOF

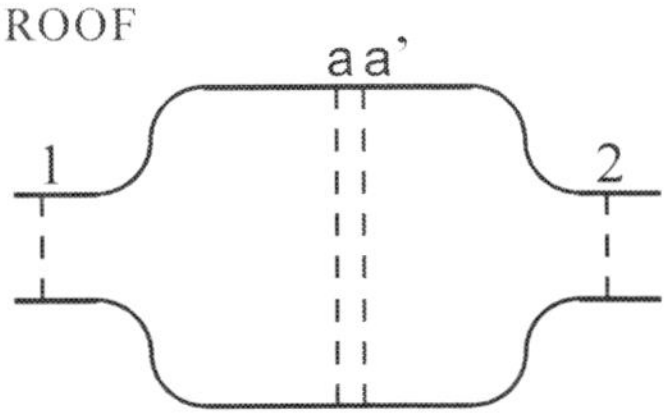

Consider an open system describing a stationary process with only one inlet in conditions 1 and one outlet in conditions 2 (see the attached schematic view). Suppose that the f uid's intensive properties are uniform along the spanwise directions at all the sections perpendicular to the streamwise direction.

When applying the balances to the part of the system between sections 1 and a, it has to be pointed out that the area of section a is now comparable to the external area of the open system, so here the quantities q and q' may not be equal in general (see discussion in section 8.3.). The correct heat that should be used in the balance is then q':

$$0 = q' - w + (h'_1 - h'_\mathrm{a})$$
$$0 = j'_s + \hat{\sigma} + (s_1 - s_\mathrm{a})$$

If now we consider the open system included between section 1 and a', infnitely close to a, we have

$$0 = q' + dq' - (w + dw) + h'_1 - (h'_\mathrm{a} + dh')$$
$$0 = j'_s + dj'_s + \hat{\sigma} + d\hat{\sigma} + s_1 - (s_\mathrm{a} + ds)$$

leading to

$$0 = dj'_s + d\hat{\sigma} - ds \Rightarrow T\,ds = dq' + T\,d\hat{\sigma} \quad ; \quad 0 = dq' - dw - dh' \Rightarrow$$
$$dw = dq' - d(e_c + e_p) - T\,ds - v\,dP = \cancel{dq'} - d(e_c + e_p) - \cancel{dq'} - T\,d\hat{\sigma} - v\,dP$$

Integrating this system from section 1 to section 2, expressions (8.15) and (8.16) are obtained.

8.8. Processes with no Work

In the particular situation of a fuid fowing through passive elements (at rest), no work is exchanged by the system. Usually, and specially in the case of gases, the potential energy can be neglected.

8.8.1. Adiabatic Flow in Ducts

The condition of adiabatic fow is satisfed in practice in many cases. Equations (8.12), (8.13) and (8.14), for a system with one input stream 1 and one output stream 2, become:

$$h'_1 - h'_2 = 0 \Rightarrow h + e_c = \text{const.} \tag{8.17}$$
$$s_2 - s_1 = \hat{\sigma} \geq 0 \tag{8.18}$$
$$e'_2 - e'_1 = -i \leq 0 \tag{8.19}$$

Lamination Valves, Nozzles, and Diffusers

All these systems are adiabatic in practice, because the fuid stays a very short time inside them and the surface of heat exchange with the outside is relatively small.

DEFINITION: A valve is a device that restricts the f ow. Throttling implies turbulent f ow, and the system is not in equilibrium in points very near it, so the process is non-static (therefore, irreversible). Throttling produces pressure losses that may be substantial, but generally the speed of f ow does not vary signif cantly.

Many times, the object of a valve is precisely to reduce pressure; a typical example of this application is found in inverse Rankine cycles (see chapter 11).

Equations (8.17), (8.18) and (8.19), applied between two sections, one previous to the valve and the other far enough downstream the valve to let the f ow recover an internal equilibrium, are simplif ed to

$$h_1 = h_2 \quad ; \quad s_1 < s_2 \quad ; \quad e_2 - e_1 = -i = -T_0(s_2 - s_1) < 0$$

DEFINITION: A nozzle is a mechanical device designed to increase the velocity of the f ow. This is achieved by changing the inner section of the duct, and implies lowering the pressure.

DEFINITION: A diffuser is a mechanical device that, changing the inner section of the duct, reduces the f uid's velocity. The aim is to increase the pressure; nevertheless, pressure might even be reduced in a diffuser in the anomalous case of an highly irreversible f ow.

Equations (8.17), (8.18) and (8.19) become:

$$h_1 + e_{c_1} = h_2 + e_{c_2} \quad ; \quad s_1 < s_2 \quad ; \quad e'_2 - e'_1 = -i = -T_0(s_2 - s_1) < 0$$

For the cases of a nozzle and a diffuser,

DEFINITION: The *isentropic eff ciency* η_s is def ned as the ratio between the specif c enthalpy change in the real process and the one that would have been attained if the process had been reversible (and thus, isentropic) going from the same initial state and ending with the same pressure. This ratio must be taken in such a way that it is smaller than one (see f gure 8.2).

In f gure 8.3 the different adiabatic f ows are shown in ducts that can be produced from a given initial state 1.

8.8.2. Bernoulli's Equation

If we have a liquid f ow inside a duct, we can consider the specif c volume practically constant across the f uid. In addition, the potential energy may play a role in this context. Equation (8.15) gives $0 = v\Delta P + \Delta e_c + \Delta e_p + \int_1^2 T\, d\hat{\sigma}$, that is

$$\frac{P_1}{\rho} + \frac{1}{2}c_1^2 + gz_1 = \frac{P_2}{\rho} + \frac{1}{2}c_2^2 + gz_2 + \int_1^2 T\, d\hat{\sigma} \tag{8.21}$$

where g is the acceleration of gravity. In the case that $\hat{\sigma}$ can be disregarded

$$\frac{P}{\rho} + \frac{1}{2}c^2 + gz = constant$$

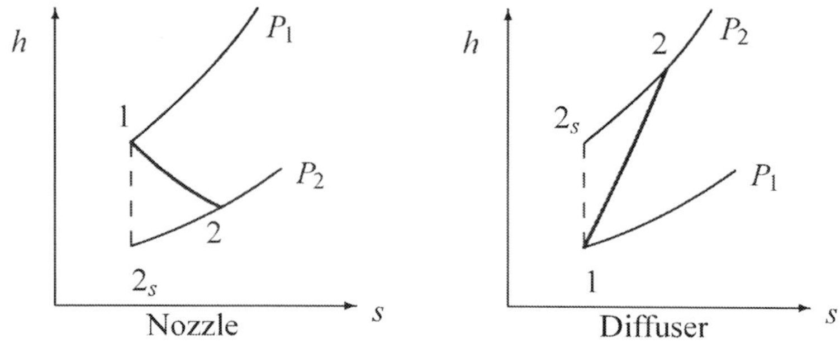

$$\text{nozzle: } \eta_s = \frac{h_1 - h_2}{h_1 - h_{2_s}} \qquad \text{diffuser: } \eta_s = \frac{h_{2_s} - h_1}{h_2 - h_1} \tag{8.20}$$

(where 2_s represents a state with pressure P_2 and entropy s_1)

Figure 8.2.

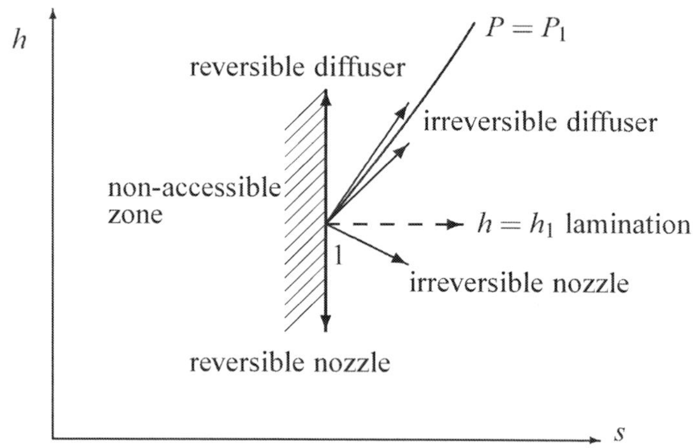

Figure 8.3. Adiabatic f ows in ducts.

8.8.3. Heat Exchanger and Boilers

In a heat exchanger, each stream describes a non-adiabatic process of movement inside passive elements ($dW = 0$; $de_p = 0$). However, the ensemble may be considered, as a f rst approximation, to be adiabatic, since heat losses to the surroundings (heat exchange through the exchanger housing) are usually very low. (For more precision, of course they can be taken into account, estimating them once the heat exchanger size and the temperature prof le along it are known approximately).

On the other hand, variations of kinetic energy are negligible in this case in practice, and therefore, the equations for the stationary process become:

$$\sum_j dm_j = 0 \quad ; \quad \sum_j h_j dm_j = 0 \quad ; \quad -\sum_j s_j dm_j = d\sigma$$

The most usual case is schematically presented in f gure 8.4; a surface heat exchanger (of the kind 'housing and tubes') with two inlets and two outlets; as there is not any mixing between both streams, chemical composition and pressure may be different for each. For that case we have:

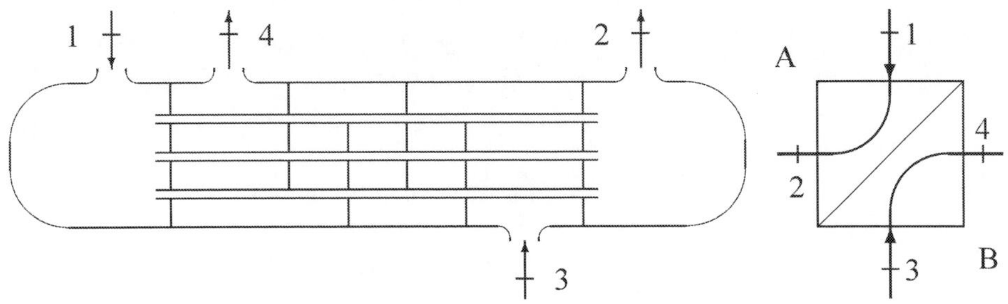

Figure 8.4. Heat exchanger.

mass balance: $\dot{m}_1 = -\dot{m}_2 \quad ; \quad \dot{m}_3 = -\dot{m}_4$ (as there is no mixing)

First Law: $\dot{m}_1 h_1 + \dot{m}_3 h_3 + \dot{m}_2 h_2 + \dot{m}_4 h_4 = 0$

Second Law: $\dot{\sigma} = -(\dot{m}_2 s_2 + \dot{m}_4 s_4 + \dot{m}_1 s_1 + \dot{m}_3 s_3)$

$$\dot{m}_1(h_1 - h_2) = \dot{m}_3(h_4 - h_3) \quad ; \quad \dot{\sigma} = \dot{m}_1(s_2 - s_1) + \dot{m}_3(s_4 - s_3)$$

For each stream, from momentum balance and Second Law:

$$-vdP = Td\hat{\sigma} \quad ; \quad \hat{\sigma}_{\mathrm{A}} = \int -\frac{v_{\mathrm{A}} dP_{\mathrm{A}}}{T_{\mathrm{A}}} \quad ; \quad \hat{\sigma}_{\mathrm{B}} = \int -\frac{v_{\mathrm{B}} dP_{\mathrm{B}}}{T_{\mathrm{B}}}$$

$$\dot{m}_1(s_2 - s_1) = \int \frac{d\dot{Q}_{\mathrm{A}}}{T_{\mathrm{A}}} + \dot{\sigma}_{\mathrm{A}} \quad ; \quad \dot{m}_3(s_4 - s_3) = \int \frac{d\dot{Q}_{\mathrm{B}}}{T_{\mathrm{B}}} + \dot{\sigma}_{\mathrm{B}}$$

If the ensemble is adiabatic, $d\dot{Q}_{\mathrm{B}} = -d\dot{Q}_{\mathrm{A}}$. Thus

$$\dot{I} = T_0(\dot{\sigma}_{\mathrm{A}} + \dot{\sigma}_{\mathrm{B}}) + T_0 \int \left(\frac{d\dot{Q}_{\mathrm{A}}}{T_{\mathrm{A}}} + \frac{d\dot{Q}_{\mathrm{B}}}{T_{\mathrm{B}}} \right)$$

$$\dot{I} = T_0 \left\{ -\int \left(\frac{\dot{V}_{\mathrm{A}} dP_{\mathrm{A}}}{T_{\mathrm{A}}} + \frac{\dot{V}_{\mathrm{B}} dP_{\mathrm{B}}}{T_{\mathrm{B}}} \right) + \int d\dot{Q}_{\mathrm{A}} \left(\frac{1}{T_{\mathrm{A}}} - \frac{1}{T_{\mathrm{B}}} \right) \right\}$$

In heat exchangers, the second integral is frequently the most important term; that is, the exergy destruction due to heat exchange through a f nite temperature step, which increases with the difference of temperatures between both streams and decreases with their product ($\frac{1}{T_A} - \frac{1}{T_B} = \frac{T_B - T_A}{T_A T_B}$). The term due to head losses can usually be neglected.

From a thermodynamic point of view, a boiler is the same as a heat exchanger, with the exception of a combustion process happening in one of the streams (chemical transformations will be studied in Chapter 10). For the other stream (water-steam), there is still a non-adiabatic process with no work.

8.9. Processes with Work

DEFINITION: A *turbine* is a device that produces work as a consequence of the expansive motion of a fuid through a system of blades inserted on a freely-rotating axis. The fuid experiences a decrease of pressure.

DEFINITION: A *compressor* is a device that performs a mechanical work on the gas that to goes through it with the aim of increasing its pressure.

DEFINITION: A *pump* is a device that carries out the same operation as the compressor, but with a liquid.

8.9.1. Adiabatic Processes

In practice, turbines, compressors and pumps' behavior is adiabatic (the heat exchange with the surroundings is minor). Therefore, the First Law is expressed in the following way

$$dW = \sum_j h'_j dm_j$$

which, in the case of only one inlet and one outlet, and considering changes on kinetic and potential energy negligible, leads to

$$w = h_1 - h_2 \tag{8.22}$$

If the process is reversible, application of the Second Law for stationary processes gives

$$0 = \sum_j s_j dm_j$$

which, in the case of only one inlet and one outlet, is just $s_2 = s_1$, or in other words, the process is isentropic.

DEFINITION: The *isentropic effciency* η_s is defned as the ratio between the specifc enthalpy change (equal to the specifc work) in the real process and the enthalpy change (or work) that would have been attained if the process had been reversible (and thus, isentropic) going from the same initial state and ending at the same pressure. This ratio must be taken in such a way that it is smaller than one (see fgure 8.5).

In the case of a pump, isobars would be very close ones to each other. The fgure would be analogous to the case of a compressor, but at a smaller scale.

8.9.2. Non-Adiabatic Processes

The most interesting case is the isothermal one (some devices with isothermal behavior do exist in practice, but only for great capacities).

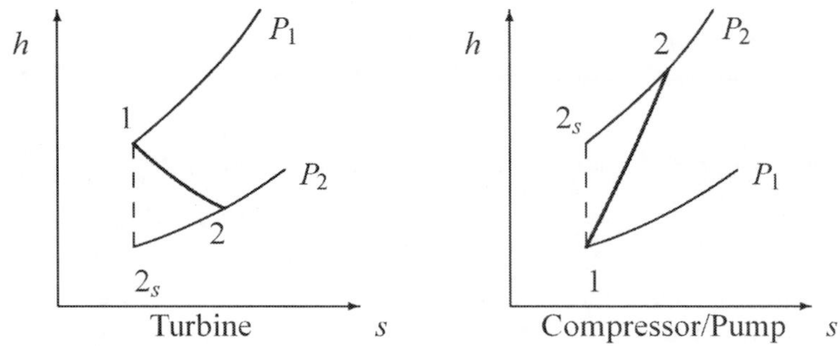

$$\text{expansion: } \eta_s = \frac{h_1 - h_2}{h_1 - h_{2_s}} = \frac{w}{w^R} \qquad \text{compression: } \eta_s = \frac{h_{2_s} - h_1}{h_2 - h_1} = \frac{w^R}{w} \tag{8.23}$$

(where 2_s represents a state with pressure P_2 and entropy s_1)

Figure 8.5. Isentropic eff ciency in processes with work.

DEFINITION: The *isothermal eff ciency* η_T is def ned as the ratio between the specif c work in the real process and in the reversible isothermal process going from the same initial state and ending in the same f nal state. This ratio must be taken in such a way that it is smaller than one.

The work in the reversible isothermal and reversible adiabatic cases, are represented graphically in f gures 8.6 or 8.7. As a consequence of the momentum balance, for a reversible

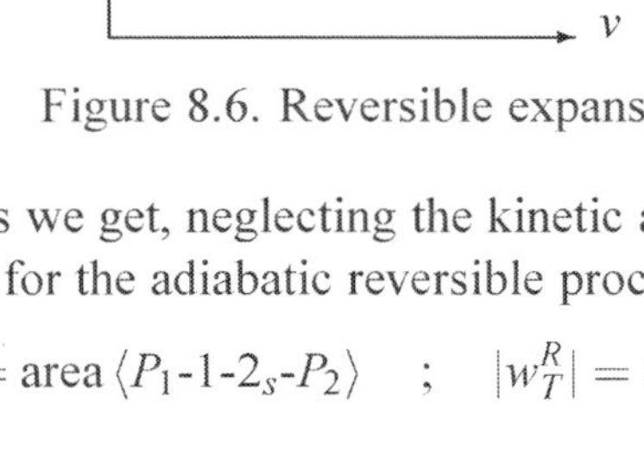

Figure 8.6. Reversible expansion.

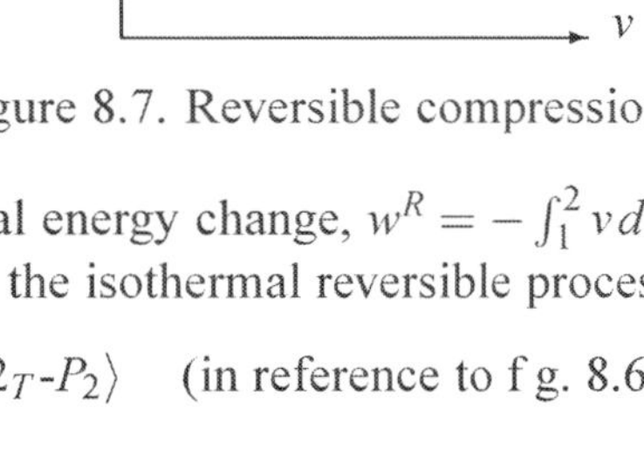

Figure 8.7. Reversible compression.

process we get, neglecting the kinetic and potential energy change, $w^R = -\int_1^2 v\,dP$; in particular for the adiabatic reversible process and for the isothermal reversible process

$|w_s^R| = \text{area}\,\langle P_1\text{-}1\text{-}2_s\text{-}P_2\rangle$; $|w_T^R| = \text{area}\,\langle P_1\text{-}1\text{-}2_T\text{-}P_2\rangle$ (in reference to f g. 8.6 and 8.7)

Since

$$\left(\frac{\partial P}{\partial v}\right)_s = \frac{\partial(Ps)}{\partial(vs)}\frac{\partial(PT)}{\partial(PT)}\frac{\partial(vT)}{\partial(vT)} = \frac{\partial(Ps)}{\partial(PT)}\frac{\partial(vT)}{\partial(vs)}\frac{\partial(PT)}{\partial(vT)} =$$

$$= \frac{T}{c_v}\frac{c_p}{T}\left(\frac{\partial P}{\partial v}\right)_T = \gamma\left(\frac{\partial P}{\partial v}\right)_T \Rightarrow \left|\left(\frac{\partial P}{\partial v}\right)_s\right| > \left|\left(\frac{\partial P}{\partial v}\right)_T\right|$$

the slope of the isentropic curve is always larger (in absolute value) than the isothermal's one at any point in a P-v diagram. Thus, for the case of a expansion we obtain

$$w_T^R > w_s^R$$

and for the case of a compression (negative work)

$$\left|w_T^R\right| < \left|w_s^R\right|$$

This justif es the interest in isothermal processes, particularly for the case of a reversible compression cooled by a f uid at ambient temperature. In practice, given the diff culties in manufacturing an isothermal turbine or compressor, a process with two adiabatic stages with an intermediate heat exchange (reheating for expansion or cooling for compression) is frequently used. Obviously, more stages could be used, leading to a further approach of the real process to the isothermal one. However, the increase of the number of stages becomes too expensive quickly.

8.9.3. Pumping liquids

For pumping liquids, the same equations as for the adiabatic compressor can be used

$$w = h_1' - h_2' \quad ; \quad w^R = -\int_1^{2_s} v\,dP - \Delta e_c - \Delta e_p \qquad (s_{2_s} = s_1) \quad ; \quad \eta_s = \frac{w^R}{w}$$

but the variations of potential energy must be taken into account in many cases, if the connection ducts between the pump and other devices are included in the system. However, the variations of kinetic energy are usually negligible. Finally, v is practically constant, so it is satisf ed

$$w^R \approx -v\Delta P - \Delta e_p \tag{8.24}$$

Exercise A power plant has two networks of vapor, one for supply at 80 bar and 400 °C and the other for consumption at 30 bar, which constantly demands 20 t/h of vapor. The second one is fed with vapor extracted from an adiabatic turbine operating at counter-pressure, taking advantage of the energetic jump between the two nets, with $\eta_{s_1} = 0.8$.

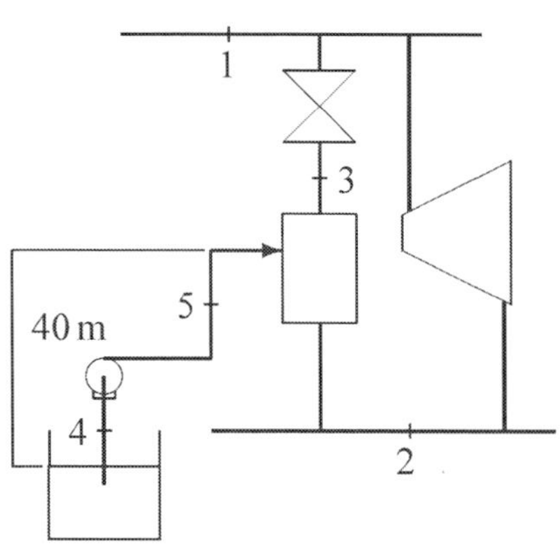

In case of mechanical failure, the supply to the low pressure net is ensured by a lamination of high-pressure vapor, reducing the resulting temperature increase by an injection of liquid water taken from a tank at 50 °C and 1 bar. The adiabatic mixing of water and laminated vapor is carried out in such a way that the same temperature of vapor in the low-pressure net is guaranteed. The injection of water is done after adiabatic pumping with $\eta_{s_2} = 0.7$ up to the pressure of the mixing chamber (the low-pressure net's one). There is a difference in height between the tank and the mixing chamber of 40 m. Calculate:

1. Powers of turbine and pump.
2. Temperature in the low-pressure net.
3. Mass f ow of water to inject.

4. Exergetic power destructions in the turbine, pump, lamination valve, and mixing chamber.

Data: Use Mollier's diagram. Neglect variations of kinetic energy in all cases, and variations of potential energy in the turbine, lamination, and mixing. For liquid water, assume that in the problem's conditions: $v = 10^{-3}\,\mathrm{m^3/kg}$; $s = 4.18\,\mathrm{kJ/kg\,K}\ln\frac{T}{273.15\,\mathrm{K}}$; $h = 4.18\,\mathrm{kJ/kg\,K}(T-273.15\,\mathrm{K}) + v(P-1\,\mathrm{bar})$. Take ambient conditions $P_0 = 1$ bar and $T_0 = 300\,\mathrm{K}$, and $\mathrm{g} = 9.8\,\mathrm{m/s^2}$.

Solution From Mollier's diagram

$$h_1 = 3\,145\,\mathrm{kJ/kg} \quad ; \quad s_1 = 6.36\,\mathrm{kJ/kg\,K}$$
$$s_{2_s} = s_1 \Rightarrow h_{2_s} = 2\,900\,\mathrm{kJ/kg} \Rightarrow h_2 = h_1 - \eta_{s_1}(h_1 - h_{2_s}) = 2\,949\,\mathrm{kJ/kg}$$

and reading from the diagram $s_2 = 6.45\,\mathrm{kJ/kg\,K}$, $t_2 \approx 280\,°\mathrm{C}$.

$$\dot{W}_\mathrm{T} = \dot{m}(h_1 - h_2) = \frac{20\,000\,\mathrm{kg}}{3\,600\,\mathrm{s}}(3\,145 - 2\,949)\,\mathrm{kJ/kg} = 1\,088.9\,\mathrm{kW}$$
$$\dot{I}_\mathrm{T} = \dot{m}T_0(s_2 - s_1) = 150\,\mathrm{kW}$$

In the lamination valve

$$h_3 = h_1 \Rightarrow s_3 = 6.78\,\mathrm{kJ/kg}$$

For point 4 (liquid):

$$h_4 \approx 4.18\,\frac{\mathrm{kJ}}{\mathrm{kg\,K}}\,50\,\mathrm{K} = 209\,\mathrm{kJ/kg}$$
$$w_\mathrm{B} = -\frac{v\Delta P + \Delta e_p}{\eta_{s_2}} = -\frac{10^{-3}(30-1)10^5\,\frac{\mathrm{J}}{\mathrm{kg}} + 40\times 9.8\,\frac{\mathrm{J}}{\mathrm{kg}}}{0.7} = -4.7\,\mathrm{kJ/kg}$$
$$h_5 = h_4 - w_\mathrm{B} - \Delta e_p = 213.3\,\mathrm{kJ/kg}$$

Applying the First Law to the mixing chamber

$$\dot{m}_5 h_5 + \left(20\,\frac{\mathrm{t}}{\mathrm{h}} - \dot{m}_5\right)h_3 = 20\,\frac{\mathrm{t}}{\mathrm{h}}\,h_2 \Rightarrow \dot{m}_5 = 20\,\frac{\mathrm{t}}{\mathrm{h}}\,\frac{h_2 - h_3}{h_5 - h_3} = 1.34\,\mathrm{t/h}$$

$$\dot{W}_\mathrm{B} = \dot{m}_5 w_\mathrm{B} = -1.75\,\mathrm{kW}$$
$$\dot{I}_L = (20 - 1.34)\,\frac{\mathrm{t}}{\mathrm{h}}\,T_0(s_3 - s_1) = 653\,\mathrm{kW}$$
$$t_5 \approx \frac{h_5 - v(P_5 - 1\,\mathrm{bar})}{4.18\,\mathrm{kJ/kg\,K}} = 50.34\,°\mathrm{C}$$
$$s_5 \approx 4.18\,\frac{\mathrm{kJ}}{\mathrm{kg\,K}}\ln\frac{323.48}{273.15} = 0.707\,0\,\mathrm{kJ/kg\,K}$$
$$s_4 \approx 4.18\,\frac{\mathrm{kJ}}{\mathrm{kg\,K}}\ln\frac{323.15}{273.15} = 0.702\,6\,\mathrm{kJ/kg\,K}$$
$$\dot{I}_\mathrm{B} = \dot{m}_5 T_0(s_5 - s_4) = 0.486\,\mathrm{kW}$$
$$\dot{I}_\mathrm{M} = T_0\dot{\sigma} = T_0(\dot{m}_2 s_2 - \dot{m}_5 s_5 - \dot{m}_3 s_3) = 126.7\,\mathrm{kW}$$

8.10. Filling and Discharging of Vessels

In this case there is a net f ow of mass going in or out of the system; hence, it is obvious that the process is non-stationary. In the usual case of a quasistatic f lling or discharging of an adiabatic, rigid vessel, we have the following balances:

$$dm = dm_1 \quad ; \quad dU' = h_1' dm_1 \quad ; \quad dS = d\sigma + s_1 dm_1$$

Let us remark that, in the case of a f lling, the state 1 is def ned by what is given upstream; in general $P_1 > P$ and $T_1 \neq T$, where P and T are variables in time. Therefore, the f lling of an adiabatic rigid vessel is generally irreversible (except in the very particular case that $T_1 = T$ and $P_1 = P$ is satisf ed at any time). However, in the quasistatic discharging there is not any cause of irreversibility in principle, so $\sigma = 0$, and the f uid goes out at every moment in the conditions inside the vessel at this instant.

If the vessel were not rigid (e.g., a vessel with a f oating ceiling), a work on the system would exist $dW = P_e dV$.

As an example, let us prove that the quasistatic and adiabatic discharging of a rigid vessel that contains a pure monophasic f uid is isentropic (specif c entropy remains constant inside the vessel): by application of the Second Law $dS = dJ_s + d\sigma + s_1 dm = s\,dm$. As it is adiabatic, $dJ_s = 0$; there are not any internal causes of irreversibility, so $d\sigma = 0$; and at every moment the f uid goes out in the conditions inside the vessel, $s_1 = s$.

$$s\,dm = dS = d(sm) = s\,dm + m\,ds \Rightarrow m\,ds = 0 \Rightarrow ds = 0 \Rightarrow s = constant$$

If the f uid inside the vessel is in a two-phase equilibrium, the specif c entropy of the outgoing stream is not equal to the specif c entropy in the vessel, but it is equal to the specif c entropy of the extracted phase. If vapor is extracted,

$$s^V dm = dS = s\,dm + m\,ds = \{xs^V + (1-x)s^L\}dm + m\,ds$$

and considering that $s^L = s^V - l/T$, we get

$$s^V dm = \left\{s^V - (1-x)\frac{l}{T}\right\} dm + m\,ds \Rightarrow m\,ds = (1-x)\frac{l}{T}dm \Rightarrow$$

$$ds = (1-x)\frac{l}{T}\frac{dm}{m}$$

and, as $(1-x)\frac{l}{Tm} > 0$ and $dm < 0$, then $ds < 0$: the specif c entropy inside the vessel decreases. It can be easily shown in a similar way that, if the extracted phase is the liquid, then the specif c entropy inside the vessel increases.

8.11. Other Non-Stationary Processes

In an open system like, for example, a turbine, the total available volume for the f uid inside the system is relatively small. In other words, the mass inside the open system at any moment is negligible; its rate of variation is also negligible.

This is true for many other cases (compressors, pumps, pipes, etc.), in which *there is no accumulation of mass, energy, entropy nor exergy*. Thus, in those cases, the terms dm, dU, and dS in the balances, referring to the variation of the corresponding property in the open system, can be neglected. Hence, for a system of this kind (with no accumulation of mass), describing a non-stationary process, equations (8.8), (8.9), (8.10), and (8.11) are still applicable.

On the other hand, the balance of momentum given in (8.15) was proved in section 8.7. for stationary processes. Nevertheless, it can be used for non-stationary processes too, considering its application to a differential of mass going through the open system at a particular instant. Indeed, the work exchanged by the differential of mass crossing the system depends on the conditions it found along its trajectory *in the particular instant it went through the system*; and not on how those conditions have been changing in time. If the conditions had been constant, the work exchanged by that differential of mass would have been the same as if they had varied later; even though it would not be the same for the next differential of mass. In conclusion, the momentum balance (8.15) can be applied to every differential of mass going through the system, obtaining the total exchanged work as

$$W = \int_{\text{start}}^{\text{end}} w\,dm$$

where w is, in general, variable along the process.

Exercise A thermally isolated vessel (A) of volume $V = 1\,\text{m}^3$ is initially full of an ideal gas of $\gamma = 1.4$, at $P_i = 20\,\text{bar}$ and $T_i = 500\,\text{K}$.

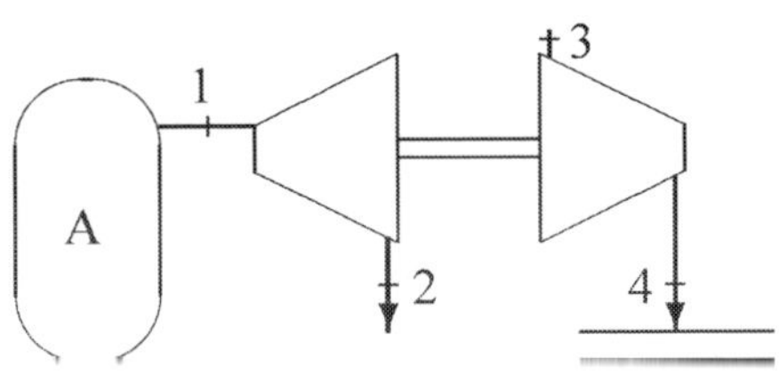

To make the most of the energy contained in the gas, the vessel is discharged through an adiabatic and reversible turbine, which activates a reversible isothermal centrifugal compressor. The compressor takes air from the environment at $P_3 = 1$ bar and $T_3 = 300\,\text{K}$ and delivers it to a network with a controlled pressure of $P_4 = 20\,\text{bar}$.

If the gas goes out from the turbine to ambient pressure (1 bar) and considering air an ideal gas, obtain the following, at the instant when turbine stops:

1. Gas temperature at turbine outlet.
2. Number of moles of gas remaining in the vessel A.
3. Total work performed by the turbine.
4. Number of moles of air introduced in the net.

Neglect all variations of kinetic and potential energy. There are no mechanical losses in the transmission of work from the turbine to the compressor.

Solution The turbine stops when pressure inside the vessel equals pressure at turbine outlet.

$$n_i = \frac{20 \cdot 10^5\,\text{Pa}\,1\,\text{m}^3}{8.314\,\frac{\text{J}}{\text{mol K}}\,500\,\text{K}} = 481.1\,\text{mol}$$

The process followed by A is adiabatic and reversible; applying the Second Law

$$dS = s_1 dn$$

The specif c entropy of the outing stream is, at any time, equal to the specif c entropy of the f uid inside the vessel at that particular moment, $s_1 = s$. Thus

$$dS = s\,dn + n\,ds = s\,dn \Rightarrow ds = 0 \Rightarrow$$

$$T_f = T_i \left(\frac{P_f}{P_i}\right)^{R/c_P} = 500\,\text{K}\left(\frac{1}{20}\right)^{\frac{0.4}{1.4}} = 212.4\,\text{K}$$

$$n_f = \frac{1 \cdot 10^5\,\text{Pa}\,1\,\text{m}^3}{8.314\,\frac{\text{J}}{\text{mol K}}\,212.4\,\text{K}} = 56.6\,\text{mol}$$

The process described by the turbine is non-stationary, since the inlet conditions (and the outlet conditions, in principle) are time dependent; however, there is not any accumulation of mass, energy, entropy nor exergy in it (at every instant, the mass contained inside the turbine is negligible), and then $dS \approx 0$. Since the process in the turbine is adiabatic and reversible, the Second Law gives $s_2 = s_1$. But, we saw that $s_1 = s$ at any time, so the specif c entropy of stream 2 is, at any time, the same as the specif c entropy of the gas inside the vessel, which in turn is constant. Given that the pressure at the turbine outlet is constant and equal to the f nal pressure in the vessel, state 2 is constant in time and coincides with the vessel's f nal state. (It is left to the reader, as an exercise, to represent in a T-s or h-s diagram what would happen if the turbine were not reversible; in this case, point 2 varies in time). Therefore $T_2 = T_f = 212.4\,\text{K}$. Applying the First Law to the ensemble turbine and vessel A:

$$dU = -dW_T + h_2 dn = dH - V\,dP \Rightarrow$$

$$W_T = h_2 \Delta n - \Delta H + V \Delta P = h_2(n_f - n_i) - (n_f h_f - n_i h_i) + V(P_f - P_i) =$$
$$= n_i(h_i - h_2) + V(P_f - P_i) = n_i \frac{\gamma}{\gamma - 1} R(T_i - T_2) + V(P_f - P_i) = 2\,126.26\,\text{kJ}$$

where integration has been performed taking into account that h_2 is constant and equal to h_f (specif c enthalpy of the gas remaining in A at the end of the process).

The process in the compressor is non-stationary, but there is not any accumulation of mass or energy in it; additionally the conditions at compressor's inlet and outlet are constant (the only thing that varies is the mass f ow). Since the process in the compressor is reversible, the momentum balance gives

$$W_C = \int_{\text{start}}^{\text{end}} w\,dn = nw = -n \int_3^4 v\,dP = -nRT_0 \ln \frac{P_4}{P_3}$$

All the work extracted by the turbine is given to the compressor, so

$$W_C = -W_T \Rightarrow n = \frac{2\,126.26\,\text{kJ}}{8.314\,\frac{\text{J}}{\text{mol K}}\,300\,\text{K}\,\ln\frac{20}{1}} = 284.6\,\text{mol}$$

Chapter 9

Multicomponent Systems

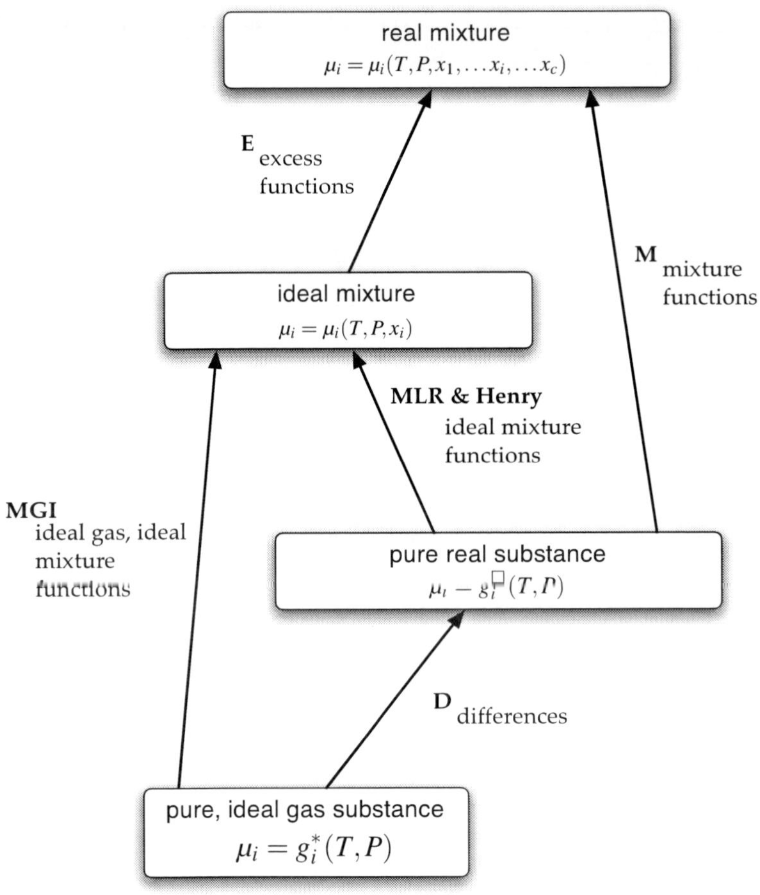

Figure 9.1. Real mixtures and ideal mixtures.

9.1. Introduction

In this chapter, we shall study some functions of state that, being only for mixtures, have not been studied in earlier chapters. These are the partial molar properties and the so-called properties of mixing. We shall extend the concepts of fugacity and coeff cient of fugacity to the components of a mixture.

We shall also introduce activity and the coeff cient of activity; two very useful properties, especially in heterogeneous or reactive systems.

As we have seen, for pure substances we have the ideal gas model, and the properties of a pure substance can be obtained adding departure functions to the properties of the ideal gas model. Similarly, for mixtures we shall def ne an ideal mixture model, and the properties of a mixture can be obtained adding excess functions to the properties of the ideal mixture model (see f gure 9.1).

9.2. Chemical Potential and Generalized Gibbs' Equations

DEFINITION: The *chemical potential* of component i in a mixture is

$$\mu_i \equiv \left(\frac{\partial U}{\partial n_i}\right)_{V,S,n_{j\neq i}} \tag{9.1}$$

For an open system with only one metric variable V, the characteristic equations (also called generalized Gibbs equations) are

$$dU = T\,dS - P\,dV + \sum_{i=1}^{C} \mu_i dn_i \tag{9.2}$$

$$dH = T\,dS + V\,dP + \sum_{i=1}^{C} \mu_i dn_i \tag{9.3}$$

$$dA = -SdT - P\,dV + \sum_{i=1}^{C} \mu_i dn_i \tag{9.4}$$

$$dG = -SdT + V\,dP + \sum_{i=1}^{C} \mu_i dn_i \tag{9.5}$$

From these equations it follows immediately that chemical potential can be expressed as:

$$\mu_i \equiv \left(\frac{\partial U}{\partial n_i}\right)_{S,V,n_{j\neq i}} = \left(\frac{\partial H}{\partial n_i}\right)_{S,P,n_{j\neq i}} = \left(\frac{\partial A}{\partial n_i}\right)_{T,V,n_{j\neq i}} = \left(\frac{\partial G}{\partial n_i}\right)_{T,P,n_{j\neq i}}$$

PROOF

For open systems the independent variables will be those of the closed system plus the amount of substance of each component, so the characteristic equation will be

$$U = U(V,S,n_1,\ldots,n_C) \tag{9.6}$$

where $n_1, \ldots, n_c$ are the amount of substance of each of the C components. Differentiating (9.6) we get

$$dU = \left(\frac{\partial U}{\partial S}\right)_{V,n_j} dS + \left(\frac{\partial U}{\partial V}\right)_{S,n_j} dV + \sum_{i=1}^{C} \left(\frac{\partial U}{\partial n_i}\right)_{V,S,n_{j\neq i}} dn_i$$

When the moles of every substance are constants, the system behaves as a closed system, therefore

$$\left(\frac{\partial U}{\partial S}\right)_{V,n_j} = T; \quad \left(\frac{\partial U}{\partial V}\right)_{S,n_j} = -P \Rightarrow dU = T\,dS - P\,dV + \sum_{i=1}^{C} \mu_i dn_i$$

Remembering the def nitions of the potentials

$$H = U + PV \quad ; \quad A = U - TS \quad ; \quad G = U + PV - TS = H - TS$$

and differentiating, the rest of the generalized Gibbs' equations are obtained immediately.

9.3. Partial Molar Properties

DEFINITION: Being Z any extensive property, it is called *partial molar property* of component i

$$\bar{\mathsf{z}}_i \equiv \left(\frac{\partial \mathsf{Z}}{\partial n_i}\right)_{P,T,n_{j\neq i}} \tag{9.7}$$

The partial molar property represents the increase in property Z when one mole of component i is added, while remaining constant pressure, temperature and the number of moles of the other components, as long as the total number of moles is big enough for the composition to remain practically constant. Any extensive property can be obtained if the corresponding partial molar properties are known, from

$$\boxed{\mathsf{Z} = \sum_{i=1}^{C} \bar{\mathsf{z}}_i n_i} \tag{9.8}$$

This important relation is a mathematical consequence of Euler's f rst theorem, because Z is an homogeneous function of degree 1 in the n_i. The partial molar property $\bar{\mathsf{z}}_i$ will be an homogeneous function of degree 0 in the n_i, therefore, must be of the form

$$\bar{\mathsf{z}}_i = \bar{\mathsf{z}}_i\left(P, T, \frac{n_1}{\sum_{i=1}^{C} n_i}, \ldots, \frac{n_C}{\sum_{i=1}^{C} n_i}\right) = \bar{\mathsf{z}}_i(P, T, x_1, \ldots, x_C)$$

where x_i are the mole fractions of the components, that must satisfy

$$\sum_{i=1}^{C} x_i = 1$$

Thus, every partial molar property is an intensive property, and depends on pressure, temperature and composition of the system.

It must be noted that, whenever the extensive variable from which it is derived has an arbitrary reference (as in the case of U or H, for example), the partial molar property will have also an arbitrary reference, and not for any other case (for example, $\overline{v_i}$ or $\overline{c_{p,i}}$).

By def nition, it follows that the chemical potential of each component is the same as its partial molar Gibbs function.

$$\mu_i = \overline{g_i} \tag{9.9}$$

This implies that the knowledge of the chemical potential of each component as a function of $P, T, n_1 \ldots n_C$, is equivalent to the knowledge of the characteristic equation of the system.

9.3.1. Gibbs-Duhem Relation

At constant P and T,

$$\left(\sum_{i=1}^{C} n_i d\mu_i = 0\right)_{P,T} \quad ; \quad \left(\sum_{i=1}^{C} x_i d\mu_i = 0\right)_{P,T} \tag{9.10}$$

By this, the variation of the chemical potential of component C can be calculated when the variation is determined for the chemical potentials of the other $C-1$ components, which is especially useful for binary systems.

PROOF

Differentiating $\mathsf{Z} = \mathsf{Z}(P, T, n_1 \ldots n_c)$ and (9.8) respectively, we obtain

$$dZ = \left(\frac{\partial Z}{\partial T}\right)_{P,n_i} dT + \left(\frac{\partial Z}{\partial P}\right)_{T,n_i} dP + \sum_{i=1}^{C} \overline{\mathsf{z}}_i dn_i \quad ; \quad dZ = \sum_{i=1}^{C} \overline{\mathsf{z}}_i dn_i + \sum_{i=1}^{C} n_i d\overline{\mathsf{z}}_i$$

and comparing both,

$$\left(\frac{\partial Z}{\partial T}\right)_{P,n_i} dT + \left(\frac{\partial Z}{\partial P}\right)_{T,n_i} dP = \sum_{i=1}^{C} n_i d\overline{\mathsf{z}}_i \tag{9.11}$$

which is called *Duhem relation*, so in a process at constant pressure and temperature must hold

$$\left(\sum_{i=1}^{C} n_i d\overline{\mathsf{z}}_i = 0\right)_{P,T} \Longrightarrow \left(\sum_{i=1}^{C} x_i d\overline{\mathsf{z}}_i = 0\right)_{P,T} \tag{9.12}$$

Writing (9.11) and (9.12) for the Gibbs function, we have

$$-SdT + VdP = \sum_{i=1}^{C} n_i d\mu_i \Longrightarrow \left(\sum_{i=1}^{C} n_i d\mu_i = 0\right)_{P,T}$$

that is the Gibbs-Duhem relation. Dividing by the total number of moles, the other cited expression is obtained.

9.3.2. Determination of Partial Molar Properties in Binary Systems

Method of the intercepts of tangent: to obtain partial molar properties of a binary mixture in a given state from experimental data, plot the mean molar property $\mathsf{z} = \mathsf{z}(x_1)$ at the pressure and temperature of interest. At the point of this curve that corresponds to the composition of interest, draw the tangent; the intercepts at $x_1 = 1$ and $x_1 = 0$ of this tangent are the partial molar property of component 1 and 2 respectively, for the given pressure, temperature, and composition.

PROOF

For a binary system, equation (9.8) can be written as $\mathsf{Z} = \bar{\mathsf{z}}_1 n_1 + \bar{\mathsf{z}}_2 n_2$ or, dividing by $n_1 + n_2$, total number of moles of the system

$$\mathsf{z} = \bar{\mathsf{z}}_1 x_1 + \bar{\mathsf{z}}_2 x_2 \tag{9.13}$$

where z is the *mean molar* property of the mixture. Differentiating this equation and combining with (9.12)

$$d\mathsf{z} = \underbrace{x_1 d\bar{\mathsf{z}}_1 + x_2 d\bar{\mathsf{z}}_2}_{0} + \bar{\mathsf{z}}_1 dx_1 + \bar{\mathsf{z}}_2 dx_2$$

and with $x_1 + x_2 = 1 \Rightarrow dx_1 + dx_2 = 0$ we get

$$d\mathsf{z} = (\bar{\mathsf{z}}_1 - \bar{\mathsf{z}}_2)dx_1 \Rightarrow \frac{d\mathsf{z}}{dx_1} = \bar{\mathsf{z}}_1 - \bar{\mathsf{z}}_2 \tag{9.14}$$

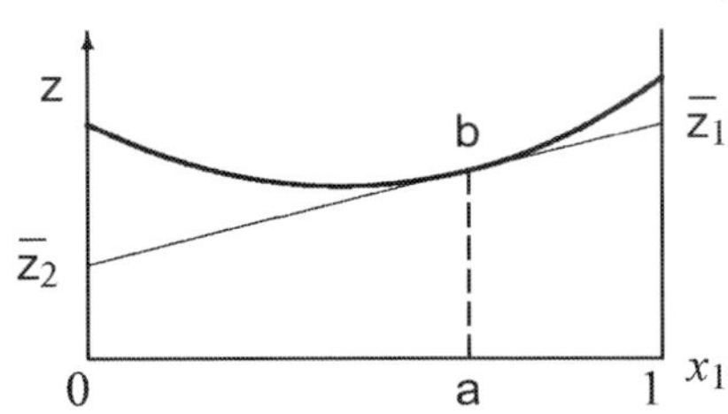

Substituting $x_2 = 1 - x_1$ in (9.13), we get the equation $\mathsf{z}(x_1) = \{\bar{\mathsf{z}}_1(x_1) - \bar{\mathsf{z}}_2(x_1)\}x_1 + \bar{\mathsf{z}}_2(x_1)$, so that its tangent in point (a,b) is (we have written z for values along the tangent, so that they can be distinguished from z, values along the curve)

$$z(x_1) - \mathsf{b} = \{\bar{\mathsf{z}}_1(\mathsf{a}) - \bar{\mathsf{z}}_2(\mathsf{a})\}(x_1 - \mathsf{a})$$
$$\mathsf{b} = z(\mathsf{a}) = \mathsf{z}(\mathsf{a}) = \{\bar{\mathsf{z}}_1(\mathsf{a}) - \bar{\mathsf{z}}_2(\mathsf{a})\}\mathsf{a} + \bar{\mathsf{z}}_2(\mathsf{a})$$

The ordinates for $x_1 = 0$ and $x_1 = 1$ for this tangent are

$$z(0) = \{\bar{\mathsf{z}}_1(\mathsf{a}) - \bar{\mathsf{z}}_2(\mathsf{a})\}(-\mathsf{a}) + \{\bar{\mathsf{z}}_1(\mathsf{a}) - \bar{\mathsf{z}}_2(\mathsf{a})\}\mathsf{a} + \bar{\mathsf{z}}_2(\mathsf{a}) \Rightarrow z(0) = \bar{\mathsf{z}}_2(\mathsf{a})$$
$$z(1) = \{\bar{\mathsf{z}}_1(\mathsf{a}) - \bar{\mathsf{z}}_2(\mathsf{a})\}\mathsf{a} + \bar{\mathsf{z}}_2(\mathsf{a}) + \{\bar{\mathsf{z}}_1(\mathsf{a}) - \bar{\mathsf{z}}_2(\mathsf{a})\}(1 - \mathsf{a}) \Rightarrow z(1) = \bar{\mathsf{z}}_1(\mathsf{a})$$

9.3.3. Thermodynamic Relations Among Partial Molar Properties

Thermodynamic relations among partial molar properties are analogous to thermodynamic relations among total properties of the system. For example, differentiating $H = U + PV$ respect to n_i at constant P, T and $n_{j \neq i}$ it is obtained:

$$\bar{h}_i = \bar{u}_i + P\bar{v}_i$$

Similarly, from $A = U - TS$ and $G = H - TS$ are obtained respectively

$$\bar{a}_i = \bar{u}_i - T\bar{s}_i \quad ; \quad \bar{g}_i = \bar{h}_i - T\bar{s}_i$$

We can say that any linear relationship among extensive properties of the open system whose coeff cients are functions of only P and T, will hold also among the corresponding partial molar properties.

Writing the condition of exact differential for each of the generalized Gibbs equations, we get the generalized Maxwell relations. There are twenty-four, for example:

$$\left(\frac{\partial \mu_i}{\partial T}\right)_{P,n_j} = -\left(\frac{\partial S}{\partial n_i}\right)_{T,P,n_{j\neq i}} = -\bar{s}_i \tag{9.15}$$

$$\left(\frac{\partial \mu_i}{\partial P}\right)_{T,n_j} = \left(\frac{\partial V}{\partial n_i}\right)_{T,P,n_{j\neq i}} = \bar{v}_i \tag{9.16}$$

Remembering that we can differentiate in any order,

$$\left(\frac{\partial \frac{\mu_i}{T}}{\partial T}\right)_{P,n_j} = \left(\frac{\partial \left(\frac{\partial \frac{G}{T}}{\partial n_i}\right)_{P,T,n_{j\neq i}}}{\partial T}\right)_{P,n_j} = \left(\frac{\partial \left(\frac{\partial \frac{G}{T}}{\partial T}\right)_{P,n_j}}{\partial n_i}\right)_{P,T,n_{j\neq i}}$$

$$= \left(\frac{\partial \frac{-H}{T^2}}{\partial n_i}\right)_{P,T,n_{j\neq i}} = -\frac{\bar{h}_i}{T^2} \tag{9.17}$$

that is called Gibbs-Helmholtz relation.

9.4. Properties of Mixing

There are a lot of well established methods for calculating properties of pure substances. Very often, these properties are used as a basis for calculating the properties of a multicomponent system. To this end, we use the so-called properties of mixing.

DEFINITION: The *property of mixing* Z^M is the difference between the value of the property Z of the mixture at P and T and the same property in the pure components of the mixture when they are pure at P and T. Denoting pure components by super-index $^\square$,

$$\boxed{\mathsf{Z}^M \equiv \mathsf{Z} - \sum_i n_i \mathsf{z}_i^\square = \sum_i n_i(\bar{\mathsf{z}}_i - \mathsf{z}_i^\square)} \tag{9.18}$$

Properties of mixing are functions of state, and depend on P, T and the number of moles of each component.

It must be noted that properties of mixing, unlike partial molar properties, *do not depend in any case* on arbitrary references, as they could always be measured from an adequate process in a closed system.

9.5. Fugacity and Fugacity Coeff cient of a Component in a Mixture

Fugacity and fugacity coeff cient for a pure substance i have been def ned through:

$$g_i^{\square} \equiv g_i^*(P^u, T) + RT \ln \frac{f_i^{\square}}{P^u} \quad ; \quad \phi_i^{\square} \equiv \frac{f_i^{\square}}{P}$$

DEFINITION: Fugacity and fugacity coeff cient for a component i in a mixture are def ned through:

$$\boxed{\mu_i \equiv g_i^*(P^u, T) + RT \ln \frac{f_i}{P^u}} \tag{9.19}$$

$$\boxed{\phi_i \equiv \frac{f_i}{x_i P}} \tag{9.20}$$

where P^u is any arbitrary pressure (its value is of no consequence).

Both fugacity f_i and fugacity coeff cient ϕ_i of a component in a mixture are intensive properties and depend on P, T, and mole fractions. From the def nition (9.19), is easy to show that

$$\left(\frac{\partial \ln f_i}{\partial P}\right)_{T, x_j} = \frac{\bar{v}_i}{RT} \quad ; \quad \left(\frac{\partial \ln f_i}{\partial T}\right)_{P, x_i} = -\frac{\bar{h}_i - h_i^*}{RT^2}$$

9.6. Activity and Activity Coeff cient

The main advantage in using fugacity instead of chemical potential is that, as can be seen from $\mu_i \equiv g_i^*(P, T) + RT \ln(f_i/P)$, for small values of P, close to zero, chemical potential has very high absolute values, hard to manipulate. We can f nd another function (activity), easier to use, replacing fugacity in many applications, if such function:

1. has a def nition independent from $g_i^*(P^u, T)$, that corresponds to the ideal gas state, hard to realize in certain substances, as for example, salts
2. is dimensionless (fugacity has the dimensions of a pressure)
3. takes values close to those of mole fraction.

DEFINITION: Activity $\mathfrak{a}_i$ and activity coeff cient γ_i for a component i in a mixture are def ned through:

$$\boxed{\mu_i(P,T) \equiv \mu_i^{\diamond}(P^{\diamond},T) + RT\,\ln\mathfrak{a}_i} \tag{9.21}$$

$$\boxed{\gamma_i \equiv \frac{\mathfrak{a}_i}{x_i}} \tag{9.22}$$

where $\mu_i^{\diamond}(P^{\diamond},T)$ is just a function of temperature T and a certain pressure $P^{\diamond}$, that can be the same than that of the mixture or a constant (usually 1 bar).

Sometimes, it will be convenient to choose $\mu_i^{\diamond}$ equal to the chemical potential of pure component i at $P^{\diamond}$, T. Other times it will not be so, but anyhow, we can *formally* identify $\mu_i^{\diamond}$ as the chemical potential of i in an hypothetical state, that we shall consider with the same state of matter (solid, liquid, or gas) of the mixture.

Evidently, γ_i will be an intensive property of the mixture and function in general of P, T, $x_1, \ldots, x_{C-1}$. Its determination is one of the main problems in Physical Chemistry because of the importance it has in physical and chemistry equilibria.

It must be remarked that in those cases where there is a practical to use fugacity, the relation between fugacity and activity is very simple:

$$\mu_i = \mu_i^{\diamond}(P^{\diamond},T) + RT\,\ln\mathfrak{a}_i = \mu_i^{*}(P^{u},T) + RT\,\ln(f_i/P^{u})$$

$$\mu_i^{\diamond} = \mu_i^{*}(P^{u},T) + RT\,\ln(f_i^{\diamond}/P^{u}) \Rightarrow$$

$$\boxed{\mathfrak{a}_i = \frac{f_i}{f_i^{\diamond}}} \tag{9.23}$$

If we choose as reference state for activity the pure component at the same pressure and temperature, $\mu_i^{\diamond} = g_i^{\square}$,

$$\mu_i = g_i^{\square} + RT\,\ln\gamma_i x_i \tag{9.24}$$

9.7. Lewis-Randall Ideal Mixture

DEFINITION: A Lewis-Randall ideal mixture is a mixture in which for every component in all the range of composition, it holds:

$$\mu_{i,\text{L-R}} \equiv g_i^{\square}(P,T) + RT\,\ln x_i \tag{9.25}$$

where superscript $^{\square}$ means pure component at the same P, T and state of matter (solid, liquid, or gas) of the mixture.

This model is a drastic simplif cation of the real behavior of a mixture, because in it the chemical potential of a component does not depend on the mole fractions of the other components, and only depends on P, T and its own mole fraction.

It can be shown by Statistical Thermodynamics reasoning, that for the model to hold, interaction between particles of kind i and i should be equal to interactions between particles of kind i and j; this happens in mixtures whose components have similar chemical structure and molecular size; therefore, its validity is quite limited, specially in liquids and solids.

From the def nition, other properties of interest can be derived:

$$\mu_{i,\text{L-R}} = g_i^*(P,T) + RT\,\ln\frac{f_{i,\text{L-R}}}{P} = g_i^*(P,T) + RT\,\ln\frac{f_i^{\square}x_i}{P} \Rightarrow \boxed{f_{i,\text{L-R}} = x_i f_i^{\square}} \tag{9.26}$$

$$\mu_{i,\text{L-R}} - g_i^{\square} = RT\,\ln x_i \Rightarrow G_{\text{L-R}}^M = RT\sum_{i=1}^{C} n_i \ln x_i \tag{9.27}$$

$$\left(\frac{\partial G_{\text{L-R}}^M}{\partial P}\right)_{T,n_j} = \boxed{V_{\text{L-R}}^M = 0} \tag{9.28}$$

$$-\left(\frac{\partial G_{\text{L-R}}^M}{\partial T}\right)_{P,n_j} = \boxed{S_{\text{L-R}}^M = -R\sum_{i=1}^{C} n_i \ln x_i} \tag{9.29}$$

$$G_{\text{L-R}}^M + TS_{\text{L-R}}^M = \boxed{H_{\text{L-R}}^M = 0} \tag{9.30}$$

$$H_{\text{L-R}}^M - PV_{\text{L-R}}^M = U_{\text{L-R}}^M = 0 \tag{9.31}$$

$$A_{\text{L-R}}^M = G_{\text{L-R}}^M + PV_{\text{L-R}}^M = G_{\text{L-R}}^M \tag{9.32}$$

$$\left(\frac{\partial H_{\text{L-R}}^M}{\partial T}\right)_{P,n_j} = \boxed{C_{p,\text{L-R}}^M = 0} \tag{9.33}$$

Differentiating (9.25), it is easy to show that:

$$\bar{v}_{i,\text{L-R}} = v_i^{\square} \quad ; \quad \bar{h}_{i,\text{L-R}} = h_i^{\square} \quad ; \quad \bar{u}_{i,\text{L-R}} = u_i^{\square} \quad ; \quad \bar{c}_{p,i,\text{L-R}} = c_{p,i}^{\square}$$

$$\bar{s}_{i,\text{L-R}} = s_i^{\square} - R\ln x_i \quad ; \quad \bar{a}_{i,\text{L-R}} = a_i^{\square} + RT\ln x_i$$

If the pure component at the same P, T and state of matter is taken as reference for activity, $f_i^{\circ} = f_i^{\square}$,

$$\boxed{a_{i,\text{L-R}} = x_i} \tag{9.34}$$

$$\boxed{\gamma_{i,\text{L-R}} = 1} \tag{9.35}$$

the activity of a component in an ideal mixture is equal to its mole fraction, and all activity coeff cients are 1.

9.7.1. Ideal Gases Mixture

DEFINITION: An ideal gases mixture is a Lewis-Randall ideal mixture in which every pure component behaves as an ideal gas.

$$\mu_{i,\text{MGI}} \equiv g_i^*(P,T) + RT\,\ln x_i \tag{9.36}$$

The ideal gases mixture model holds in practice when density is very low. In this case, all interactions between particles can be neglected, so the Lewis-Randall model will hold. For example, air, across a wide range of pressure and temperature, behaves as an ideal gases mixture.

Comparing (9.36) and fugacity def nition (9.19), for an ideal gases mixture

$$f_{i,\text{MGI}} = x_i P \quad ; \quad \phi_{i,\text{MGI}} = 1 \tag{9.37}$$

Remembering (9.28), in an ideal gases mixture

$$V = \sum_{i=1}^{C} n_i v_i^* + \cancelto{0}{V_{\text{L-R}}^M} = n_T \frac{RT}{P} \Rightarrow PV = n_T RT$$

where n_T is the total number of moles. This is the thermal equation of the mixture, formally identical to that of a pure ideal gas.

If we def ne the *partial pressure* P_i of a component i in a mixture as the pressure n_i mole of the pure component with the same V and T of the mixture, in this model

$$P_i V = n_i RT \Rightarrow P_i = x_i P = f_{i,\text{MGI}} \tag{9.38}$$

which is known as Dalton's law.

Exercise 1 1 mol of ideal gas A, with $c^*_{pA} = 30\,\text{J/mol K}$, at $T_{A_1} = 400\,\text{K}$ and $P_1 = 1$ bar, is mixed adiabatically with 3 mol of another ideal gas B, with $c^*_{pB} = 38\,\text{J/mol K}$ at $T_{B_1} = 300\,\text{K}$ and $P_1 = 1$ bar, keeping pressure constant. Calculate the temperature of the mixture, once the equilibrium is reached, and the variation of entropy in the process.

Solution: It is a closed system that performs an isobaric with the ambient process, therefore

$$Q = \Delta H = 0 = \cancelto{0}{H^M} + \sum_i n_i c^*_{p_i} (T_{i_1} - T_2) \Rightarrow$$

$$T_2 = \frac{\sum_i n_i c^*_{p_i} T_{i_1}}{\sum_i n_i c^*_{p_i}} = \frac{1\,\text{mol}\; 30\,\frac{\text{J}}{\text{mol K}}\; 400\,\text{K} + 3\,\text{mol}\; 38\,\frac{\text{J}}{\text{mol K}}\; 300\,\text{K}}{1\,\text{mol}\,30\,\frac{\text{J}}{\text{mol K}} + 3\,\text{mol}\; 38\,\frac{\text{J}}{\text{mol K}}} = 320.8\,\text{K}$$

As pressure is constant,

$$\Delta S = S^M + \sum_i n_i c^*_{p_i} \ln \frac{T_2}{T_{i_1}} = -8.314\,\frac{\text{J}}{\text{mol K}} \left(1\,\text{mol}\,\ln\frac{1}{4} + 3\,\text{mol}\,\ln\frac{3}{4} \right) +$$
$$+ \left(1\,\text{mol}\; 30\,\frac{\text{J}}{\text{mol K}} \ln\frac{320.8}{400} + 3\,\text{mol}\; 38\,\frac{\text{J}}{\text{mol K}} \ln\frac{320.8}{300} \right) = 19.7\,\frac{\text{J}}{\text{K}}$$

Exercise 2 An adiabatic mixing chamber is in stationary state, with an inlet f ow (1) of 3 mol/s of ideal gas A and another (2) of 9 mol/s of ideal gas B, both at 25 °C and 1.1 bar. The only outlet f ow (3) is at 1 bar. Calculate the outlet temperature and the entropy generation in the chamber per second.

Solution: It is a stationary non-work adiabatic process:

$$dU' = dQ - dW + \sum_j h'_j dm_j = 0 \Rightarrow h_3\dot{m}_3 = h_2\dot{m}_2 + h_1\dot{m}_1$$

$$h_3\dot{m}_3 = h^{\square}_{\mathrm{A}}(P_3,T_3)\dot{m}_1 + h^{\square}_{\mathrm{B}}(P_3,T_3)\dot{m}_2 + \cancelto{0}{H^M_3}$$

As A and B are ideal gases, their enthalpies does not depend on pressure, but only in T, then

$$h^{\square}_{\mathrm{A}}(T_3)\dot{m}_1 + h^{\square}_{\mathrm{B}}(T_3)\dot{m}_2 = h^{\square}_{\mathrm{B}}(T_2)\dot{m}_2 + h^{\square}_{\mathrm{A}}(T_1)\dot{m}_1$$

and with $T_2 = T_1 \Rightarrow T_3 = T_1 = 25\,°\mathrm{C}$

$$dS = dJ_s + d\sigma + \sum_j s_j dm_j \Rightarrow \dot{\sigma} = \dot{m}_3 s_3 - \dot{m}_2 s_2 - \dot{m}_1 s_1$$

$$\dot{m}_3 s_3 = \dot{m}_2 s^{\square}_{\mathrm{B}}(P_3,T_3) + \dot{m}_1 s^{\square}_{\mathrm{A}}(P_3,T_3) - \sum_i \dot{m}_i R \ln x_i \Rightarrow$$

$$\dot{\sigma} = -\dot{m}_2 R \ln\frac{P_3}{P_2} - \dot{m}_1 R \ln\frac{P_3}{P_1} - R\left(\dot{m}_1 \ln\frac{\dot{m}_1}{\dot{m}_1+\dot{m}_2} + \dot{m}_2 \ln\frac{\dot{m}_2}{\dot{m}_1+\dot{m}_2}\right) =$$

$$= 8.314\,\frac{\mathrm{J}}{\mathrm{mol\,K}}\left(12\,\frac{\mathrm{mol}}{\mathrm{s}}\ln\frac{1.1}{1} - 3\,\frac{\mathrm{mol}}{\mathrm{s}}\ln\frac{3}{12} - 9\,\frac{\mathrm{mol}}{\mathrm{s}}\ln\frac{9}{12}\right) = 65.62\,\frac{\mathrm{W}}{\mathrm{K}}$$

9.8. Excess Function

DEFINITION: Excess function is the difference between the value of the property in the mixture and the value the same property would have if the mixture had been a Lewis-Randall ideal mixture.

Relating both values with mixing functions, it is seen that excess function can be expressed also as the difference between the real and the Lewis-Randall ideal mixture mixing properties, and therefore, it does not depend on any arbitrary reference

$$\boxed{\mathsf{Z}^E \equiv \mathsf{Z} - \mathsf{Z}_{\text{L-R}} = \mathsf{Z}^M - \mathsf{Z}^M_{\text{L-R}}} \tag{9.39}$$

For example, excess Gibbs function is

$$G^E = G^M - RT\sum_{i=1}^{C} n_i \ln x_i$$

and if we take as reference state for activity the pure components at the same P, T and state of matter of the mixture $\mu_i^{\diamond} = \mu_i^{\square}$, we have

$$G^M = \sum_{i=1}^{C}\mu_i n_i - \sum_{i=1}^{C}\mu_i^{\square} n_i = \sum_{i=1}^{C}(\mu_i - \mu_i^{\square})n_i = \sum_{i=1}^{C}(\mu_i - \mu_i^{\diamond})n_i = RT\sum_{i=1}^{C} n_i \ln a_i$$

$$G^E = RT\sum_{i=1}^{C} n_i \ln a_i - RT\sum_{i=1}^{C} n_i \ln x_i \Rightarrow \boxed{G^E = RT\sum_{i=1}^{C} n_i \ln\gamma_i} \tag{9.40}$$

Notice that the knowledge of $g_i^{\square}(T,P)$ for each component and $G^E(T,P,n_1\ldots n_c)$ or, the similarities of what is the same, for all the $\gamma_i(T,P,x_1\ldots n_{C-1})$, is equivalent to the knowledge in the characteristic equation of the mixture.

As for the excess function difference, it permutes with differentiation; so, the following relations are immediate

$$V^E = \left(\frac{\partial G^E}{\partial P}\right)_{T,n_j} = V^M \quad ; \quad S^E = -\left(\frac{\partial G^E}{\partial T}\right)_{P,n_j} = S^M + R\sum_{i=1}^{C} n_i \ln x_i$$
$$H^E = G^E + TS^E = H^M \quad ; \quad U^E = H^E - PV^E = U^M$$
$$A^E = G^E + PV^E = A^M - RT\sum_{i=1}^{C} n_i \ln x_i \quad ; \quad C_p^E = \left(\frac{\partial H^E}{\partial T}\right)_{P,n_j} = C_p^M$$

Applying Gibbs-Duhem relation to (9.40), the following important relation can be proved:

$$\boxed{RT\ln\gamma_i = \left(\frac{\partial G^E}{\partial n_i}\right)_{T,P,n_{j\neq i}}} \tag{9.41}$$

Exercise 3 Two equal volumes V_0 of two liquids A and B, initially in mechanical and thermal equilibrium with the ambient at T_0, P_0, are mixed adiabatically and at constant pressure. The mixture results at $T_1 < T_0$ and it occupies a volume V_1. This mixture is heated up to T_0, keeping it at ambient pressure, and it is measured that in this process it absorbs heat Q and it ends with a volume V_2. Knowing these experimental data, obtain: H^M, V^M, U^M, H^E, V^E, U^E of the mixture at P_0, T_0 and with the composition of the experience.

Solution By def nition of mixing property $V^M = V_2 - 2V_0$. As the whole process is isobaric with the ambient $\Delta H_{0-2} = Q$ and as $\Delta H_{0-2} = H^M = Q$,

$$U^M = H^M - P_0V^M = Q - P_0(V_2 - 2V_0); \qquad H^E = H^M - H^M_{L-R} = H^M = Q$$
$$V^E = V^M - V^M_{L-R} = V^M = V_2 - 2V_0; \qquad U^E = H^E - P_0V^E = Q - P_0(V_2 - 2V_0)$$

Exercise 4 An adiabatic vessel in stationary state has an inlet f ow $\dot{n}_1 = 1$ mol/s of pure component 1, at P and $t_1 = 50\,°\mathrm{C}$, and another inlet f ow $\dot{n}_2 = 2$ mol/s of pure component 2 at P and $t_2 = 70\,°\mathrm{C}$, and one outlet f ow of the mixture at P and t_3. The mixture at P has $g^E = 1\,000x_1x_2$ J/mol, that we shall assume independent of T in the conditions of interest. The pure component isobaric specif c heats are $c_{p1}^{\square} = 50$ J/mol K and $c_{p2}^{\square} = 70$ J/mol K. Calculate t_3 and the entropy generation per second in this process.

Solution From the fourth Gibbs relation, is immediate that

$$s^E = -\left(\frac{\partial g^E}{\partial T}\right)_{P,x_1} = 0 \Rightarrow \begin{cases} h^E = g^E = h^M \\ s^M = s^E + s^M_{L-R} = s^M_{L-R} = -R\sum_{i=1}^{2} x_i \ln x_i \end{cases}$$

$$\dot{H}_1 + \dot{H}_2 = \dot{H}_3 \Rightarrow \dot{n}_1 h_1^{\square}(P,t_1) + \dot{n}_2 h_2^{\square}(P,t_2) = \dot{n}_1 h_1^{\square}(P,t_3) + \dot{n}_2 h_2^{\square}(P,t_3) + \dot{H}_3^E \Rightarrow$$

$$t_3 = \frac{\dot{n}_1 c_{p1}^{\square} t_1 + \dot{n}_2 c_{p2}^{\square} t_2 - 1\,000\,\frac{\text{J}}{\text{mol}}\,\frac{\dot{n}_1\dot{n}_2}{\dot{n}_1+\dot{n}_2}}{\dot{n}_1 c_{p1}^{\square} + \dot{n}_2 c_{p2}^{\square}} = 61.23\,^{\circ}\text{C}$$

$$\dot{\sigma} = \dot{n}_1 c_{p1}^{\square} \ln\frac{T_3}{T_1} + \dot{n}_2 c_{p2}^{\square} \ln\frac{T_3}{T_2} - R\left(\dot{n}_1 \ln\frac{1}{3} + \dot{n}_2 \ln\frac{2}{3}\right) = 13.96\ \text{W/K}$$

9.9. Very Dilute Solution (Henry's Model)

DEFINITION: Calling D to solvent and S to solute (there might be several) Henry's ideal mixture is def ned by

$$\mu_S = \mu_S^{\square'}(P,T) + RT\ \ln x_S \qquad (9.42)$$

The state of the solute denoted by super-index $^{\square'}$ is not any of the real solute; it even depends on the solvent. Nevertheless, as the Gibbs function of a pure substance is a characteristic equation, $\mu_S^{\square'}(P,T)$ def nes thermodynamically these hypothetical states of the f uid. Taking as reference for activity this hypothetical state, we have $\alpha_S = x_S$ and $\gamma_S = 1$. The fugacity in the hypothetical state $f_S^{\square'}$, given by

$$\mu_S^{\square'} = g_S^*(T,P^u) + RT\ln\frac{f_S^{\square'}}{P^u}$$

is called *Henry's constant*, a non-zero f nite value, and from the def nition of fugacity,

$$\mu_S = g_S^*(T,P^u) + RT\ln\frac{f_S}{x_S P^u} \Rightarrow \boxed{f_S = x_S f_S^{\square'}(P,T)} \qquad (9.43)$$

Behavior in the limit 3 *(part 1)* ***For a component in a mixture:***
When the mole fraction of a component in a mixture tends to 1, this component follows the Lewis-Randall model.

Behavior in the limit 3 *(part 2)* ***For a component in a mixture:***
When the mole fraction of a component in mixture tends to 0, this component follows the Henry's model.

Equation (9.43) has been found experimentally to be obeyed by every non-ionic solutions in the limit $x_S \to 0$, so it is very useful for the study of dilute solutions. Henry's model is applied frequently to liquids solutions, in which pressure affects only very slightly to other thermodynamical properties. Therefore, in such solutions, Henry's constant is almost invariant with pressure.

It can be shown, applying Gibbs-Duhem relation, that when the solute S follows Henry's model, the solvent D follows Lewis-Randall model

$$\boxed{\mu_D = \mu_D^{\square}(P,T) + RT\ \ln x_D} \tag{9.44}$$

Figure 9.2 illustrates graphically the behavior of a binary non-ionic mixture homogeneous

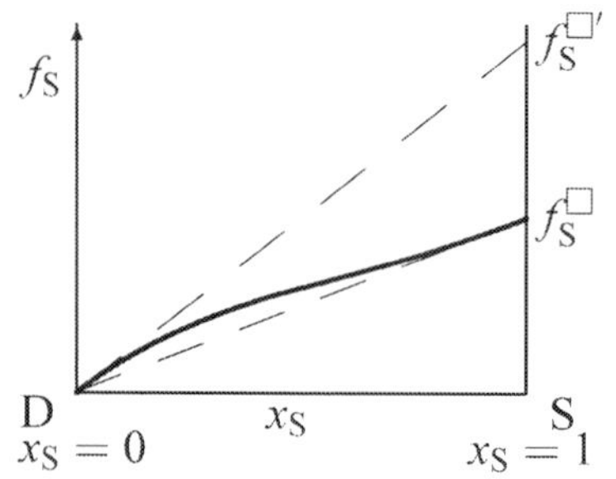

Figure 9.2.

in all the range of composition, kept at constant pressure and temperature. Notice that at $x \to 0$ Henry's model (9.42) holds, and at $x \to 1$, holds Lewis-Randall (9.25).

9.10. Equilibrium Conditions

We shall consider only liquid-vapor equilibrium of a mixture in a closed system with only V as metric variable, and we shall neglect the effect of any external f eld and the interface energy; under these restrictions, the equilibrium conditions are:

$$T^L = T^V \tag{9.45}$$

$$P^L = P^V \tag{9.46}$$

$$\forall i: \quad \mu_i^L = \mu_i^V \Rightarrow f_i^L = f_i^V \tag{9.47}$$

PROOF

As we saw in the study of phase equilibria for pure substances, the two f rst conditions correspond to thermal and mechanical equilibrium under the given assumptions. These kinds of equilibrium are obviously necessary for thermodynamical equilibrium. To obtain the last condition, we must consider that both phases (liquid and vapor) are open systems, whose characteristic equation will be given by (9.2), and that U, S, V, and n_i are additives

($dn_i^L = -dn_i^V$, because the whole system is closed and there is no chemical reaction)

$$dU = dU^V + dU^L = T\,dS^V - P\,dV^V + \sum_i \mu_i^V\,dn_i^V + T\,dS^L - P\,dV^L + \sum_i \mu_i^L\,dn_i^L$$
$$= T(dS^V + dS^L) - P(dV^V + dV^L) + \sum_i (\mu_i^V\,dn_i^V + \mu_i^L\,dn_i^L)$$
$$= T\,dS - P\,dV + \sum_i (\mu_i^V - \mu_i^L)dn_i^V \Rightarrow dS = \frac{dU}{T} + \frac{P}{T}dV - \frac{\Sigma_i(\mu_i^V - \mu_i^L)dn_i^V}{T}$$

Keeping the whole system closed, $dU = 0$, $dV = 0$, $dS \geq 0$. If the system was initially in equilibrium, its entropy cannot change, because if it changes, the system could not return to its original state, so necessarily $dS = 0$ in equilibrium, therefore $\mu_i^V = \mu_i^L$.

This reasoning could easily be generalized to any other phase equilibrium with any number of phases in a multicomponent system.

9.10.1. Practical Equilibrium Formulation

For a gas phase, properties of the mixture can be obtained from the same thermal equations of state that have been used for pure substances, using empiric mixing rules that give mixture parameters as a combination of those of the pure components; the simplest one is Kay's rule, that takes for the mixture just a mean value, weighted with the mole fractions. For example, Lee-Kesler tables can be used for estimating compressibility factors, departure functions and fugacity coeff cients for a gaseous mixture, calculating reduced values dividing pressure and temperature by pseudo-critical properties:

$$T_{cm} = \sum_i x_i T_{ci} \quad ; \quad P_{cm} = \sum_i x_i P_{ci}$$

There are also more complex mixing rules that are more accurate, but only for the gaseous phase. All these methods fails to give enough accuracy for the liquid phase; in this case, activity must be used:

$$\forall i: \qquad f_i^L \equiv a_i^L f_i^{\square L} \equiv \gamma_i^L x_i f_i^{\square L} \tag{9.48}$$

where γ_i^L is the activity coeff cient of each component in the liquid phase; $f_i^{\square L}$ is the fugacity in the state reference for activity, for which we shall take always the pure component in liquid state at P and T of the system under equilibrium; and x_i is the mole fraction of each component in the liquid phase.

Notice that for the relatively volatile components (of higher vapor pressure) the stable state at P and T of the system might be vapor instead of liquid, for this 'hypothetical'; nevertheless, even in this case $f_i^{\square L}$ will be a well-def ned function, that can be obtained from experimental data or by extrapolation of the stable liquid fugacity to the hypothetical states.

When $f_i^{\square L}$ corresponds to a stable liquid, its determination is immediate, resulting, as we have seen in (7.12)

$$f_i^{\square L} = \phi_i^{\square V s} P_i^s \exp\left(\frac{v_i^{\square L}\{P - P_i^s\}}{RT}\right) = \Pi P_i^s \tag{9.49}$$

where superscript s stands for 'saturated'. *Poynting's correction* Π is generally close to 1 at low or moderate pressures P, as the reader can deduce. Thus, equations (9.47) can be written as:

$$\boxed{\forall i: \quad \gamma_i^L x_i P_i^s \Pi_i = \phi_i^V y_i P} \tag{9.50}$$

with the caveat that reference liquid states for activity should be stable; in another case, $P_i^s \Pi_i$ must be substituted by the function $f_i^{\square L}$ corresponding to the hypothetical states. It is interesting to note that activity coeff cients vary only very slightly with pressure, if the system is far from its critical state, and that in practice can be considered as depending only in temperature and composition.

Equations (9.50) solve in theory the practical equilibrium problems; we shall mention that for its frequency, the bubble point (for a liquid at a given pressure, f nd the temperature at which the f rst bubble of vapor appears, and the composition of this vapor), and the dew point (for a vapor at a given pressure, f nd the temperature at which the f rst drop of liquid appears and the composition of this liquid).

9.10.2. Simple Cases

1. Equilibrium at low pressures.
   If pressure is now enough and the nature of components permits it, vapor phase can be considered as an ideal gas mixture, and Poynting correction to be 1. Then,

$$\gamma_i^L x_i P_i^s = y_i P \tag{9.51}$$

2. Liquid phase as an ideal Lewis-Randall mixture.
   If the nature of the components makes it possible to consider liquid phase as a Lewis-Randall ideal mixture, and also at a vapor phase,

$$x_i P_i^s \Pi_i = \phi_i^{\square V} y_i P \tag{9.52}$$

   Where the fugacity coeff cient in the vapor phase corresponds now to the pure component instead of to the component in the mixture.

3. Low pressure and ideal Lewis-Randall mixture.
   If both simplif cations are valid,

$$\boxed{x_i P_i^s = y_i P} \tag{9.53}$$

   That is the well-known Raoult's Law. Adding the C equations (9.53) we have $\sum x_i P_i^s = P$ that gives the vapor pressure of a mixture of a given composition.

Exercise 5 A mixture of butane (component 1) and propane (component 2) in liquid-vapor equilibrium is contained in a vessel. The mole fraction of butane in the liquid is $x = 0.9$. Find the pressure in the vessel and the composition of the vapor phase. Assume that Raoult's law is accurate enough for this mixture. Vapor pressure of pure propane at 20 °C is $P_2^s = 8.39$ bar and for the pure butane at the same temperature, $P_1^s = 2.08$ bar.

Solution: Applying Raoult's law, being x and y the mole fractions of butane in liquid and vapor respectively,

$$\left.\begin{array}{r} xP_1^s = yP \\ (1-x)P_2^s = (1-y)P \end{array}\right\} \Rightarrow \left\{\begin{array}{l} P = (0.9 \times 2.08 + 0.1 \times 8.39)\,\text{bar} = 2.71\,\text{bar} \\ y = \frac{0.9\times 2.08}{2.71} = 0.69 \end{array}\right.$$

Exercise 6 In a closed space at 1 bar and 300 K, there are 400 m^3 of air in equilibrium with water. Air can be considered as an ideal gas mixture. Specifc volume of liquid water is $v^L = 1.003 \cdot 10^{-3}\,\text{m}^3/\text{kg}$, constant along the isotherm in all conditions considered. Vapor pressure of water is given by

$$\ln P = 14.03 - \frac{5210}{T}$$

giving T in K and P in bar. For water $M = 18$ g/mol. Neglect the solubility of air in water. Find the mass (kg) of water in the air.

Solution: For water, the equilibrium condition is $f_{H_2O}^L = f_{H_2O}^V = x_{H_2O}P$. As we neglect solubility of air in water, liquid phase can be considered as pure water; therefore

$$\left(\frac{\partial \ln f}{\partial P}\right)_T = \frac{v}{RT} \Rightarrow \ln \frac{f_{H_2O}^L(1\,\text{bar}, 300\,\text{K})}{f_{H_2O}^L(P^{\text{sat}}, 300\,\text{K})} = \int_{P^{\text{sat}}}^{1\,\text{bar}} \frac{v^L}{RT}dP = \frac{v^L}{RT}(P - P^{\text{sat}})$$

$$P^{\text{sat}} = \exp\left(14.03 - \frac{5210}{300}\right)\,\text{bar} = 0.0356\,\text{bar} \Rightarrow$$

$$f_{H_2O}^L(1\,\text{bar}, 300\,\text{K}) = f_{H_2O}^L(P^{\text{sat}}, 300\,\text{K}) \exp\left\{\frac{v^L(P - P^{\text{sat}})}{RT}\right\} =$$

$$= f_{H_2O}^{V□}(P^{\text{sat}}, 300\,\text{K}) \times 1.0395 = 1.0395\,P^{\text{sat}} = 0.03696\,\text{bar} \quad ; \quad x_{H_2O} = 0.03696$$

$$\sum_i n_i = \frac{PV}{RT} = 16036\,\text{mol} \Longrightarrow n_{H_2O} = 592.7\,\text{mol} = 10.7\,\text{kg}$$

Chapter 10

Reactive Systems

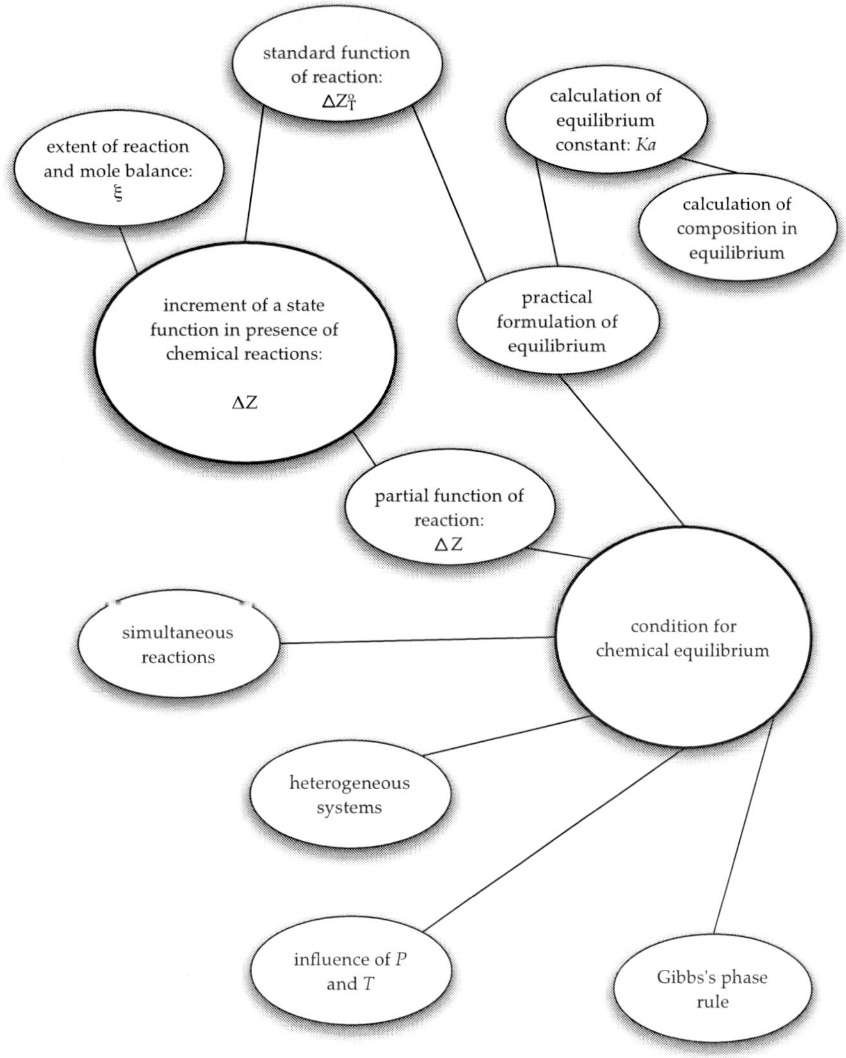

10.1. Introduction

Why should we talk at all about reactive systems?

After all, engineers need Thermodynamics for three reasons: writing energy balances (First Law), studying eff ciencies (Second Law), and calculating thermodynamic properties. The expressions for the First and Second Laws, as it has been pointed out, are valid for reactive systems, the only restriction of their validity being the relativistic effects (nuclear reactions, for example). Nothing else needs to be said about the calculation of thermodynamic properties.

By def nition, a state property depends only on the state of the system, not in the way the system has reached that state. So, the change of any property of a reactive system between two given states (enthalpy, entropy, etc.) would be exactly the same if, instead, the system were not reactive at all but it exchanged the appropriate amounts of mass of each component with the surroundings, so that it reached the same f nal state.

In spite of this, there are three reasons why we need to study reactive systems explicitly:

1. First and Second Laws *don't give a physical meaning to U (or H) nor S*, but *they do for their variations in a closed system*. Therefore, as some components might be converted into others through chemical reactions in a closed system, *it isn't correct to take references for each component independently*, as is usually done when tabulating thermodynamic properties. Should compatible references for enthalpy of every component had been chosen, there would not have been a need for anything more to write to correct balances in reactive systems. Applying the Third Law, we know that a suitable reference for every pure substance (crystalline and in internal equilibrium), whether it is simple or a compound, is at 0 K, because in the limit of zero temperature the entropy change in a chemical reaction is zero. Therefore, for entropy it would have been enough to take references in that state. But, unfortunately, references for h and s have been taken in a incompatible way, so now we need to study how to correct them, which we shall accomplish through the so-called standard functions of reaction.

2. Also, many times it is necessary to consider certain non-equilibrium states in reactive systems. For example, a combustible-air mixture in near ambient conditions, very often can be for a long time in what may look like an equilibrium state; however, it can be shown that it is not just by adding a convenient catalyst to speed up the reaction, so that changes in the state variables will become observable.

   We shall treat only those non-equilibrium states in which the only missing equilibrium condition is the chemical equilibrium. Such states will be def ned by the same variables that def ne true equilibrium states, plus one or more variables that determine the actual composition, and how much it differs from the equilibrium one. These variables are the *extent* of the possible reactions.

3. We shall also study the chemical equilibrium condition, because of its great practical interest.

This chapter is a starting point for an entire line of thermodynamics focused on chemistry. It is therefore natural that some preparatory concepts are developed summarizing the basic conceptual and mathematical tools that will be needed. However, the standard engineer will not require specialized training in the subtleties of the discipline, so the chapter will soon evolve towards the real problems: f rst, how to calculate the increment of a state function (internal energy, entropy) which will allow to calculate the heat and work involved in a process where chemical reactions take place; second, how to calculate the properties of a system in which there exists a chemical equilibrium, and how this is changed by changes in T,P and the composition of the system.

10.2. Extent of Reaction and Mole Balance

We shall represent a chemical reaction as

$$\sum_{i=1}^{C} \nu_i \mathrm{A}_i = 0 \tag{10.1}$$

where ν_i is the stoichiometric coeff cient for component A_i, whose sign is positive if it is a product and negative if it is a reactant, and A_i is the formula of component i in the conventional writing.

DEFINITION: *The extent of reaction ξ is the number of moles of component i formed in the reaction, divided by the stoichiometric coeff cient of the component in the reaction*

By def nition, the extent of reaction is measured in mole. Notice that it is not needed to specify which component is taken to calculate the extent of reaction, because all the moles formed will be proportional to the stoichiometric coeff cients, and the quotient will be the same for any component (the consumption of a mole is mathematically the same as -1 mol formed).

For a closed system, with only one reaction and which initially has a composition given by n_i^0,

$$n_i = n_i^0 + \nu_i \xi \Rightarrow \xi = \frac{n_i - n_i^0}{\nu_i} \tag{10.2}$$

$$dn_i = \nu_i d\xi \tag{10.3}$$

If there are several reactions, the number of moles of component i formed will be the sum of those formed by each reaction. We shall take into account that the component's stoichiometric coeff cient will be different in each reaction, and that each reaction will have its own extent of reaction. For a system with R reactions, the mole balance for each component will be

$$\boxed{n_i = n_i^0 + \sum_{k=1}^{R} \nu_i^k \xi_k} \tag{10.4}$$

For open systems, the balance will be the same, except that n_i^0 will no longer be the initial number of moles (a constant therefore), but the balance of the initial number and the moles of the component given to the system or taken out of it by the inlet and outlet streams, thus, a function of time.

10.3. Partial Function of Reaction

In the following, we shall consider closed systems with one phase, in thermal and mechanical equilibrium but not in chemical equilibrium. As the easiest to measure variables that are enough to def ne such states are P, T and composition, it is convenient to use them as independent variables to express the rest of the functions. Often, we shall f nd the derivative of some extensive function Z with respect to ξ, at constant P and T, that we shall call *partial function* Z *of reaction.*

DEFINITION: Partial function of reaction is the derivative of the function with respect to extent of reaction, at constant P and T and in a closed system.

$$\triangle \mathsf{Z} \equiv \left(\frac{\partial \mathsf{Z}}{\partial \xi}\right)_{P,T} = \sum_{i=1}^{C} \left(\frac{\partial \mathsf{Z}}{\partial n_i}\right)_{P,T,n_{j\neq i}} \frac{dn_i}{d\xi} = \sum_{i=1}^{C} \nu_i \bar{\mathsf{z}}_i \tag{10.5}$$

by (10.3). Notice than even when some partial molar functions (as partial molar enthalpy) depend on an arbitrary reference, partial functions of reaction could be measured through a process in a closed system, and therefore never depend on any reference. That implies that for the given equation to hold, care should be taken to have chosen compatible references in partial molar functions.

Exercise 1 Calculate $\triangle V$ for the ideal gases reaction at 1 bar and 298 K

$$2N_2 + O_2 \rightarrow 2N_2O$$

Solution As it is an ideal gases mixture

$$\triangle V(1\,\text{bar}, 298\,\text{K}) = \sum_i \nu_i \bar{v}_i(1\,\text{bar}, 298\,\text{K}) = \sum_i \nu_i \frac{R\,298\,\text{K}}{1\,\text{bar}} = -0.0248\,\text{m}^3/\text{mol}$$

10.4. Standard Function of Reaction

As it has been mentioned, a major issue with reactive systems is to correct the references of some thermodynamic funcions (energy, enthalpy, entropy, etc.), that have been taken in an arbitrary way for each component. What we need for this is to quantify the change in the desired function for the pure components, as if the reaction took place under specif c conditions, which can be chosen by convention.

For this, by analogy with the partial function of reaction, we shall def ne the *standard function* Z *of reaction*:

DEFINITION: Standard function Z of reaction is

$$\boxed{\triangle \mathsf{Z}_T^\circ = \sum_{i=1}^{C} \nu_i \mathsf{z}_i^\circ} \tag{10.6}$$

where z_i° is the specif c function of the pure component in standard conditions: P° (in old tables usually 1 atm; in modern ones, 1 bar) and temperature T.

Matter is usually assumed as solid or liquid after what the real state of the component is in these conditions. Ideal gas is assumed for those components that are gases in the same conditions. In any case, the table should specify which state is tabulated (often, tables include also hypothetical states other than ideal gas such as 'inf nite dilution' for example).

The temperature of standard conditions is not def ned (while P° is). As we shall see later, it is enough to know standard functions of reaction at just one temperature, so usually only the values at 25 °C are tabulated.

Notice that it is not necessary to know standard function of reaction for variables whose partial molar function does not depend on arbitrary reference. Therefore, there is no need for a correction when using these references, as is the case for volume or c_p. Neither is it needed (although it may be used) for entropy, because, applying the Third Law, it is enough to have tabulated absolute entropies from 0 K at one temperature. Therefore, only standard enthalpies of reaction (needed to correct enthalpy references in First Law) and standard Gibbs function of reaction (needed for chemical equilibrium studies) are necessary.

It is not needed to know the standard function of reaction for every reaction of interest. As we are considering state functions, they can be calculated considering *any* process provided it develops between the same initial and f nal states as the studied reaction.

Any linear combination of reactions that add up to the reaction of interest, allows us to calculate the standard function of the reaction of interest, just as the same linear combination of the standard functions of reaction ('Hess law').

For example, to calculate $\triangle \mathsf{Z}^\circ_{298}$ from standard functions of formation,

$$\triangle \mathsf{Z}^\circ_{298} = \sum_{i=1}^{C} \nu_i \triangle_f \mathsf{Z}^\circ{}_{i,298} \tag{10.7}$$

where $\triangle_f \mathsf{Z}^\circ{}_{i,298}$ are the standard function of formation (at 25 °C) for every compound in the reaction, the reaction of formation being always the reaction of formation of one mole of the compound from elements.

Exercise 2 For a certain ideal gases reaction without change in the number of moles, $\triangle G^\circ$ has been calculated for a certain range of temperatures from equilibrium compositions. Results are well represented by the equation

$$\triangle G^\circ = \mathsf{A} + \mathsf{B}T + \mathsf{C}T^2$$

where A, B and C are known constants. Calculate, as a function of T: $\triangle V^\circ$, $\triangle H^\circ$, $\triangle S^\circ$, $\triangle U^\circ$, and $\triangle C_p{}^\circ$ for that range of temperature. Standard state is pure components at 1 bar.

Solution As operator $\triangle^\circ$ permutes with derivative, it is immediate that

$$\triangle S^\circ = \left(\frac{\partial \triangle G^\circ}{\partial T}\right)_P = -(\mathsf{B}+2\mathsf{C}T) \quad ; \quad \triangle H^\circ = \triangle G^\circ + T\triangle S^\circ = \mathsf{A} - \mathsf{C}T^2$$

$$\triangle V^\circ = \sum \nu_i \frac{RT}{P} = 0 \quad ; \quad \triangle U^\circ = \triangle H^\circ - P\triangle V^\circ = \triangle H^\circ$$

$$\triangle C_p{}^\circ = T\left(\frac{\partial \triangle S^\circ}{\partial T}\right)_P = -2\mathsf{C}T$$

10.5. Computing the Change of a Function $\Delta\mathsf{Z}$ in a Reactive System

In theory, it could be calculated by

$$\Delta\mathsf{Z} = \int_1^2 \left(\frac{\partial \mathsf{Z}}{\partial T}\right)_{P,\xi} dT + \left(\frac{\partial \mathsf{Z}}{\partial P}\right)_{T,\xi} dP + \Delta\mathsf{Z}\, d\xi$$

but in practice the derivatives needed are not usually known.

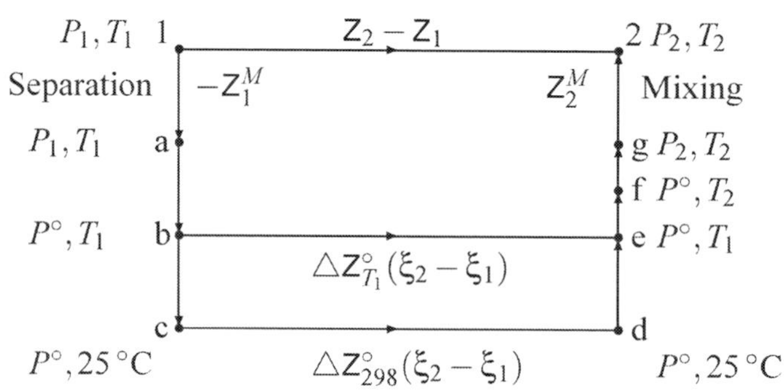

Figure 10.1. Practical computation of $\Delta\mathsf{Z}$

However, being Z a state function, its increment between 1 and 2 can be calculated along any process. We shall arbitrarily choose one, whose steps we know how to compute. For example, in a closed system (see f gure 10.1):

1-a : Separation at P_1, T_1 of the initial state: $\Delta\mathsf{Z}_{1\text{-a}} = -\mathsf{Z}_1^M$

a-c : Change of Z for pure components in the initial amounts, from P_1, T_1 to the values in thermochemical tables (P° and 25 °C)

c-d : Change of composition (pure components at P° and 25 °C) to f nal state composition. Notice that for every substance, the specif c value of the property considered is the same in points 'c' and 'd', and therefore, every component present in the system that does not appear in the reaction will be canceled out in making the difference. The change in the property will be given by the specif c value for each substance, multiplied by the number of moles formed by the reaction between states 'c' and 'd', that is: $\Delta\mathsf{Z}_{\text{c-d}} = \Delta\mathsf{Z}^\circ_{298}(\xi_2 - \xi_1)$

d-g : Change of Z for pure components in the f nal state amounts when changing P and T from P° and 25 °C to P_2, T_2

g-2 : Mixing of the components at constants P_2, T_2: $\Delta Z_{\text{g-s}} = Z_2^M$

The f gure also sketches how to calculate $\triangle Z_T^\circ$ from $\triangle Z_{298}^\circ$, and some alternative ways of computing ΔZ.

> Although the scheme in 10.1 is quite frequently useful, it is only a particular case; but it gives clear insight on how to treat other cases, for example, a furnace with separate inlets of air and combustible.

10.6. Computing $\triangle Z_T^\circ$

For **volume**, as we have already said, nothing is necessary, as volume does not depend on any reference.

For **enthalpy**:

1. Calculate $\triangle H_{298}^\circ$ from tables of standard enthalpy for formation or combustion (unfortunately, usually called in tables erroneously *heat* instead of enthalpy)
2. Calculate $\triangle H^\circ$ at any T from

$$\triangle H_T^\circ = \triangle H_{298}^\circ + \int_{298.15\,\text{K}}^{T} \left(\frac{\partial \triangle H_T^\circ}{\partial T}\right)_P dT = \triangle H_{298}^\circ + \int_{298.15\,\text{K}}^{T} \triangle C_{p}{}_T^\circ dT$$

as $\triangle^\circ$ permutes with derivative. Care should be taken that no pure components change its state of matter (solid, liquid, or gas) from 298 K to T, because in this case this equation should be corrected with the corresponding latent heats of change of state.

In practice, for combustibles are primarily used:

Higher heating value $HHV = -\triangle H_{298}^\circ$ for the combustion reaction of one mole of combustible, with water (combustion product) in its liquid state at 25 °C and 1 bar.

Lower heating value LHV: same as HHV but with ideal gas state as reference for water; the difference between both values is very nearly approximated by the latent heat of vaporization of water at 25 °C multiplied by the number of moles of water formed in the combustion of one mole of combustible.

Exercise 3 Calculate the heat absorbed when 2.5 mol of propane at external pressure of 1.2 atm and 800 K dehydrogenate in 40% following reaction:

$$CH_3{-}CH_2{-}CH_3\ (g) \rightarrow CH_3{-}CH{=}CH_2\ (g) + H_2\ (g)$$

Products of reaction stay at the same pressure and temperature. The reactive mixture can be considered as an ideal gases mixture.

Solution As reaction begins with pure propane, we follow this process:
-step 1: change of pure propane to pure f nal components, at the same P and T; the change in enthalpy is $\triangle H^\circ_{800}\ \Delta\xi$
-step 2: mixing pure f nal components, at constants P and T. As the system is an ideal gases mixture, the change in enthalpy in this step is null.

1. We calculate $\triangle H^\circ_{298}$

   (a) From tables that list standard heats of formation: (Note: many tables used the symbol $\triangle H^\circ_f$ for standard heat of formation.)
   $\triangle_f H^\circ_{298\,C_3H_6} = 4.88$ kcal/mol; $\triangle_f H^\circ_{298\,C_3H_8} = -24.82$ kcal/mol and naturally $\triangle_f H^\circ_{298\,H_2} = 0$. In our case:

   $$\triangle H^\circ_{298} = \triangle_f H^\circ_{298\,C_3H_6} + \triangle_f H^\circ_{298\,H_2} - \triangle_f H^\circ_{298\,C_3H_8} = 29.7\,\text{kcal/mol}$$

   (b) From tables of heat of combustion at 1 atm and 25 °C

   $$\begin{array}{rl} C_3H_8\ (g) + 5O_2\ (g) \rightarrow 3CO_2\ (g) + 4H_2O\ (l) & \triangle H^\circ = -530.6\,\text{kcal/mol} \\ 3CO_2\ (g) + 3H_2O\ (l) \rightarrow C_3H_6\ (g) + \frac{9}{2}O_2\ (g) & \triangle H^\circ = 492.0\,\text{kcal/mol} \\ H_2O\ (l) \rightarrow H_2\ (g) + \frac{1}{2}O_2\ (g) & \triangle H^\circ = 68.3\,\text{kcal/mol} \\ \hline C_3H_8\ (g) \rightarrow C_3H_6\ (g) + H_2\ (g) & \triangle H^\circ = 29.7\,\text{kcal/mol} \end{array}$$

   Thus $\triangle H^\circ_{298} = 29.7\,\text{kcal/mol} = 124.3\,\text{kJ/mol}$ for the given reaction.

2. We calculate $\triangle H^\circ_{800}$: $\triangle H^\circ_{800} = \triangle H^\circ_{298} + \int_{298.15\,K}^{800\,K} \triangle C_{pT}^\circ dT$; $\triangle C_{pT}^\circ = \Sigma_i \nu_i c^*_{p,i}$ and from c^*_p tables in $\frac{\text{cal}}{\text{mol K}}$

   $$\begin{array}{rl} c^*_{p\,C_3H_6} &= 0.753 + 5.691\cdot 10^{-2}\ T - 2.910\cdot 10^{-5}\ T^2 + 5.880\cdot 10^{-9}\ T^3 \\ c^*_{p\,H_2} &= 6.952 - 0.04576\cdot 10^{-2}\ T + 0.09563\cdot 10^{-5}\ T^2 - 0.2079\cdot 10^{-9}\ T^3 \\ c^*_{p\,C_3H_8} &= -0.966 + 7.279\cdot 10^{-2}\ T - 3.755\cdot 10^{-5}\ T^2 + 7.580\cdot 10^{-9}\ T^3 \\ \hline \triangle C^\circ_{pT} &= 8.671 - 1.63376\cdot 10^{-2}\ T + 0.94063\cdot 10^{-5}\ T^2 - 1.9079\cdot 10^{-9}\ T^3 \end{array}$$

   $$\triangle H^\circ_{800} = \triangle H^\circ_{298} + \int_{298.15\,K}^{800\,K} (8.671 - 1.63376\cdot 10^{-2}\ T + 0.94063\cdot 10^{-5}\ T^2 -$$

   $$-1.9079\cdot 10^{-9}\ T^3)dT = 29700\,\frac{\text{cal}}{\text{mol}} + 1181\,\frac{\text{cal}}{\text{mol}} = 30881\,\text{cal/mol} = 129.2\,\text{kJ/mol}$$

   None of the reacting gases changes to liquid in the range 298 K to 800 K, at 1 atm, so the calculation of $\triangle H^\circ_{800}$ has been carried out correctly.

3. We calculate the heat absorbed in reaction. Applying the First Law to the closed system:

   $$Q = H_2 - H_1 = \triangle H^\circ_{800}(\xi_2 - \xi_1) = 129.2\,\frac{\text{kJ}}{\text{mol}}\,2.5\,\text{mol}\,(0.4 - 0) = 129.2\,\text{kJ}$$

For **entropy**:

1. Calculate $\triangle S^\circ_{298}$ from absolute entropy tables at 25 °C and 1 atm (alternatively, the same method as for enthalpy could be used, if tables for standard entropy of formation are available)
2. Calculate $\triangle S^\circ$ at other T from

$$\triangle S^\circ_T \equiv \triangle S^\circ_{298} + \int_{298.15\,\mathrm{K}}^{T} \left(\frac{\partial \triangle S^\circ_T}{\partial T}\right)_P dT = \triangle S^\circ_{298} + \int_{298.15\,\mathrm{K}}^{T} \frac{\triangle C_{PT}^\circ}{T} dT$$

in case that no component change its state of matter in the range from 298.15 K to T.

For **Gibbs function** use the same method as for enthalpy or others:

1. Using tables of standard Gibbs functions

$$\triangle G^\circ_{298} \equiv \sum_{i=1}^{C} \nu_i \triangle_f G^\circ_{298\,i}$$

$$\triangle G^\circ_T \equiv \triangle G^\circ_{298} - \int_{298\,\mathrm{K}}^{T} \triangle S^\circ_T dT$$

2. From $\triangle H^\circ_T$ and $\triangle S^\circ_T$,

$$\triangle G^\circ_T = \triangle H^\circ_T - T \triangle S^\circ_T$$

3. There are also other methods, as for example, using Planck function and standard enthalpy of formation at 0 K. This method is usually more accurate than the one we have explained before, as the error in determining $\triangle S^\circ$ is multiplied by T, but we shall not explain it here.

10.7. Chemical Equilibrium Condition

The chemical equilibrium condition for a system with R reactions is

$$\forall k: \qquad \triangle G_k = \sum_{i=1}^{C} \nu_i^k \mu_i = 0 \tag{10.8}$$

PROOF

To f nd the chemical equilibrium condition, we shall consider an homogeneous closed system with volume as only a metric variable. As the mass of every component might change in the reaction, this must be taken into account. The characteristic equation will be the same as for an open system, in which there were inlet and outlet f ows and no reactions, provided that initial and f nal states were the same in both cases. This would lead to the same change in U:

$$dU = T\,dS - P\,dV + \sum_{i=1}^{C} \mu_i dn_i \Rightarrow dS = \frac{dU}{T} + \frac{P}{T} dV + \frac{\sum_{i=1}^{C} \mu_i dn_i}{T}$$

Each dn_i is a function of $d\xi$, as we saw in (10.3),

$$dS = \frac{dU}{T} + \frac{P}{T}dV + \frac{\sum_{i=1}^{C} \mu_i \nu_i d\xi}{T}$$

If the system is isolated, $dV = 0$, $dU = 0$, and $dS \geq 0$; if $dS > 0$ the system cannot return to its initial state, so it will not be in equilibrium; therefore, in equilibrium $dS = 0$, and

$$\sum_{i=1}^{C} \mu_i \nu_i = \sum_{i=1}^{C} \bar{g}_i \nu_i = \Delta G = 0$$

10.7.1. Practical Equation of Chemical Equilibrium

If the reference state for activity is the same as the reference state for standard function of reaction, in equilibrium

$$\boxed{\ln \prod_{i=1}^{C} a_i^{\nu_i} = \ln K_a = -\frac{\Delta G_T^\circ}{RT}} \tag{10.9}$$

PROOF

From (10.8) applied to a system with only one reaction, for the sake of simplicity, and putting chemical potentials as a function of activity

$$0 = \sum_{i=1}^{C} \nu_i \mu_i = \sum_{i=1}^{C} \nu_i (\mu_i^\diamond + RT \ln a_i) = \sum_{i=1}^{C} \nu_i \mu_i^\diamond + \sum_{i=1}^{C} RT \nu_i \ln a_i = \Delta G^\circ + RT \ln \prod_{i=1}^{C} a_i^{\nu_i}$$

provided that the reference state for activity is the same as the reference state for standard functions of reaction, $\mu_i^\circ = \mu_i^\diamond$. The expression $\prod_{i=1}^{C} a_i^{\nu_i}$ is called *chemical equilibrium constant* and it is represented by K_a; similarly, $K_z = \prod_{i=1}^{C} \left(\frac{z_i}{z^\circ}\right)^{\nu_i}$ is used for any variable z_i, where z° is a unity constant with the same dimensions; for example, for concentration $c^\circ = 1$ mol/l.

It must be remarked that $\Delta G_T^\circ = \sum_{i=1}^{C} \nu_i \mu_i^\circ$ and therefore, equilibrium constant depends on:

1. The set of stoichiometric coeff cients ν_i chosen for the reaction. If all stoichiometric coeff cients are multiplied by the same number, the reaction is still balanced, but K_a changes.
2. The reference state for μ_i°, that must be the same as the reference state for activity in each reactive component. Usually, pure component at T of the system and 1 bar, in the stable state of matter under such conditions is taken, except if it is gas, when ideal gas state is usually taken. So, generally $\mu_i^\circ = g_i^\square(P^\circ, T)$.

In solution reactions, the reference state is often the hypothetical state at inf nite dilution.

10.7.2. Calculation of Equilibrium Constant

To obtain K_a there are several methods:

1. From $\triangle G_T^\circ$, calculated by any of the earlier mentioned methods, applying (10.9)
2. From equilibrium constant at another temperature.
   If K_a is known at temperature T_1, at another temperature T:

$$\left(\frac{\partial \ln K_a}{\partial T}\right)_P = \left(\frac{\partial\left(-\frac{\triangle G_T^\circ}{RT}\right)}{\partial T}\right)_P = \frac{\triangle H_T^\circ}{RT^2} \tag{10.10}$$

   called Van't Hoff relation. Finally

$$\ln\frac{K_a(T_2)}{K_a(T_1)} = \int_{T_1}^{T_2} \frac{\triangle H_T^\circ}{RT^2} dT \approx \frac{\triangle H_{T_1}^\circ}{R}\left(\frac{1}{T_1} - \frac{1}{T_2}\right)$$

   if the range of temperatures is small enough to consider $\triangle H^\circ$ constant.

Exercise 4 Find equilibrium constant at 800 K and 1 atm for the reaction of the exercise in page 147.

Solution $\triangle G_{800}^\circ = \triangle H_{800}^\circ - 800\,\mathrm{K}\,\triangle S_{800}^\circ$. From the exercise in page 147: $\triangle H_{800}^\circ =$ 129.2 kJ/mol. From absolute entropy tables:

$$\begin{aligned}
\triangle S_{298}^\circ &= (63.8 + 31.211 - 64.51)\,\mathrm{cal/mol\,K} = 30.501\,\mathrm{cal/mol\,K} \\
\triangle S_{800}^\circ &= \int_{298\,\mathrm{K}}^{800\,\mathrm{K}} \left(\frac{8.671}{T} - 1.63376\cdot 10^{-2} + 9.4063\cdot 10^{-6}\,T - 1.9079\cdot 10^{-9}\,T^2\right) dT \\
&\quad + 30.501\,\frac{\mathrm{cal}}{\mathrm{mol\,K}} = 30.501\,\frac{\mathrm{cal}}{\mathrm{mol\,K}} + 2.645\,\frac{\mathrm{cal}}{\mathrm{mol\,K}} = 138.68\,\frac{\mathrm{J}}{\mathrm{mol\,K}} \Rightarrow \\
\triangle G_{800}^\circ &= (129.2 - 800\times 0.13868)\,\mathrm{kJ/mol} = 18259\,\mathrm{J/mol} \Rightarrow \\
K_a &= \exp\left(-\frac{18259}{8.314\times 800}\right) = 0.0642
\end{aligned}$$

10.7.3. Calculation of Equilibrium Composition

The most frequently found cases are:
Equilibrium at known P and T (isothermal-isobaric reactor).
For each component, activity at known P and T will be the only function in the composition of the system, that is, the extent of reaction from an initial known composition. Therefore, knowing equilibrium constant at given P and T

$$K_a = \prod_{i=1}^{C} a_i^{\nu_i}$$

is enough to f nd the composition of the system in equilibrium.

Exercise 5 A system consists initially of pure propane. Find its degree of decomposition to propane in equilibrium at 800 K and 1 atm. Assuming $\triangle H^{\circ} \approx 129.2$ kJ/mol constant from 800 K to 900 K, f nd the degree of decomposition at 900 K and 1 atm.

Solution Considering initially 1 mol of propane:

	C_3H_8 (g)	$\rightleftharpoons$	C_3H_6	+	H_2	total
initial moles	1		0		0	
equilibrium moles	$1-x$		x		x	$1+x$

Taking as reference state pure gases at 800 K and 1 atm: $\mathfrak{a}_i = \frac{f_i}{f_i^{\circ}} = \frac{f_i}{1\,\text{atm}}$ and remembering that it is an ideal gas mixture: $f_i = P_i = y_i P \Rightarrow \mathfrak{a}_i = y_i P/1$ atm $= y_i$ because system is at $P = 1$ atm.

$$K_{\mathfrak{a}} = \frac{\mathfrak{a}_{C_3H_6}\mathfrak{a}_{H_2}}{\mathfrak{a}_{C_3H_8}} = \frac{y_{C_3H_6}y_{H_2}}{y_{C_3H_8}}$$

From the exercise in page 151, $K_{\mathfrak{a}} = 0.0643$; and $y_i = n_i/n_t$, therefore, in equilibrium:

$$0.0643 = \frac{\frac{x}{1+x}\frac{x}{1+x}}{\frac{1-x}{1+x}} = \frac{x^2}{1-x^2}$$

that gives $x = 0.246$, that is, in equilibrium 24.6% of the initial moles will have decomposed. Applying van't Hoff relation

$$\ln\frac{K_{\mathfrak{a}}(900\,\text{K})}{K_{\mathfrak{a}}(800\,\text{K})} = \int_{800\,\text{K}}^{900\,\text{K}} \frac{\Delta H^{\circ}}{RT^2}dT \approx \frac{\Delta H^{\circ}_{800}}{R}\left(\frac{1}{800\,\text{K}} - \frac{1}{900\,\text{K}}\right) = 2.158 \Rightarrow$$

$$K_{\mathfrak{a}}(900\,\text{K}) = 0.5566 = \frac{x^2}{1-x^2} \Longrightarrow x = 0.598$$

Equilibrium at given P and without heat exchange during reaction (isobaric-adiabatic reactor).

Energy balance and chemical equilibrium condition form a system of two equations and two unknowns.

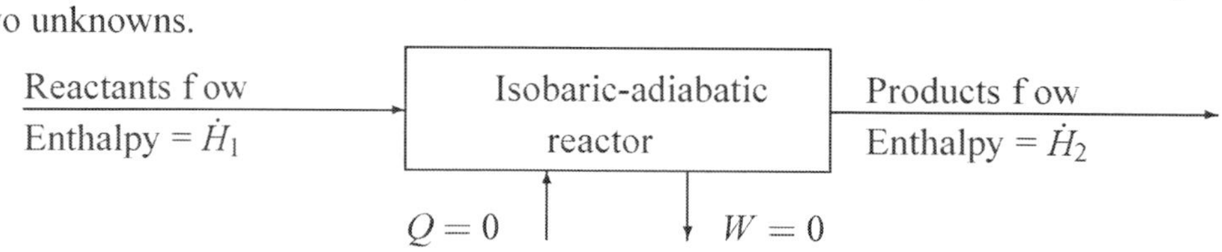

$$\dot{H}_1 = \dot{H}_2$$

First Law:
$$\sum_{i=1}^{C}(n_i^0 + \nu_i\xi)\overline{h}_i(P, T_{\text{f nal}}) = \sum_{i=1}^{C} n_i^0\overline{h}_i(P, T_{\text{initial}})$$

Second Law:
$$\prod_{i=1}^{C}\mathfrak{a}_i^{\nu_i} \text{ in equilibrium } = K_{\mathfrak{a}} \text{ at } T_{\text{f nal}}.$$

When the equilibrium constant of the reaction is very high, f nal composition can be calculated directly, because the reaction will be complete. That is usually the case for combustion reactions. The temperature obtained in this case is called *adiabatic f ame temperature*. Sometimes, the calculation of adiabatic f ame temperature may be incorrect if:

1. Specif c heat used are not valid in all the range of temperatures from T_{initial} to $T_{\text{f nal}}$
2. Final temperature is so high that products may dissociate, which means, that there are more reactions to be considered than those initially supposed.

10.7.4. Inf uence of P and T in the Equilibrium Extent of Reaction

In a system at equilibrium, the equilibrium shifts to counteract any externally imposed change (Le Chatelier's principle):

1. The extent of reaction in equilibrium at constants P and T increases with T when reaction is endothermic, and decreases when is exothermic.
2. The extent of reaction in equilibrium at constants P and T increases with P when reaction reduces volume, and decreases when produces an expansion.

PROOF

As

$$\left(\frac{\partial\left(\frac{\Delta G}{T}\right)}{\partial T}\right)_{P,\xi} = -\frac{\Delta H}{T^2} \quad ; \quad \left(\frac{\partial\left(\frac{\Delta G}{T}\right)}{\partial P}\right)_{T,\xi} = \frac{\Delta V}{T}$$

$$\left(\frac{\partial\left(\frac{\Delta G}{T}\right)}{\partial \xi}\right)_{P,T} = \frac{1}{T}\left(\frac{\partial^2 G}{\partial \xi^2}\right)_{P,T} \equiv \frac{1}{T}\Delta^2 G$$

we can write

$$d\left(\frac{\Delta G}{T}\right) = -\frac{\Delta H}{T^2}dT + \frac{\Delta V}{T}dP + \frac{\Delta^2 G}{T}d\xi$$

For a system in equilibrium, $\Delta G = 0$ always, and if equilibrium is stable, $\left(\frac{\partial^2 G}{\partial \xi^2}\right)_{P,T} = \Delta^2 G$ will always be positive. Therefore

$$0 = -\frac{\Delta H}{T^2}dT + \frac{\Delta V}{T}dP + \frac{\Delta^2 G}{T}d\xi$$

Keeping constant P and changing T in a closed system

$$\left(\frac{\partial \xi}{\partial T}\right)_P = \frac{\Delta H}{T\,\Delta^2 G} \qquad \text{con} \qquad \Delta^2 G > 0$$

the extent of reaction increases with T for an endothermic reaction in equilibrium, at constant P.

If the system is an ideal gas mixture, as $\Delta H = \Delta H^\circ$, it will be enough to investigate the sign of the standard enthalpy of reaction at the temperature of interest.

Similarly

$$\left(\frac{\partial \xi}{\partial P}\right)_T = -\frac{\Delta V}{\Delta^2 G} \qquad \text{con} \qquad \Delta^2 G > 0$$

and the extent of reaction in equilibrium increases with P when reaction reduce volume at constant T.

If the system is an ideal gas mixture, $\triangle V = \sum_i \nu_i RT/P$, and the sign of the partial volume of reaction will be the same of $\sum_i \nu_i$: if pressure increases, the equilibrium extent of reaction will increase if $\sum_{i=1}^{C} \nu_i$ is negative. If it is zero, pressure will have no effect in equilibrium extent of reaction.

If the system is an ideal gas mixture, also:

1. If an inert gas is added, every mole fraction will be lowered in the same proportion, and the effect will be the same as that of a decrease in pressure.
2. Taking as reference state pure gases in its ideal state at P° ($f_i^\circ = P^\circ$)

$$K_a = K_f = K_x \left(\frac{P}{P^\circ}\right)^{\sum_{i=1}^{C} \nu_i}$$

$$K_x = \prod_{i=1}^{C} x_i^{\nu_i} = \frac{\prod_{\text{products}} x_i^{\nu_i}}{\prod_{\text{reactants}} x_i^{-\nu_i}} = K_a \left(\frac{P}{P^\circ}\right)^{-\sum_{i=1}^{C} \nu_i}$$

So, if more reactants are added to the system, the equilibrium extent of reaction will increase.

Exercise 6 Gibbs function of formation for ethane and ethylene at 1 bar are

$$\triangle_f G^\circ{}_{C_2H_6} = -24900\,\frac{\text{cal}}{\text{mol}} + 51\,\frac{\text{cal}}{\text{mol K}}\,T \quad ; \quad \triangle_f G^\circ{}_{C_2H_4} = 9100\,\frac{\text{cal}}{\text{mol}} + 19\,\frac{\text{cal}}{\text{mol K}}\,T$$

1. Calculate the degree of decomposition of ethane at 700 K and 1 bar, by reaction

$$C_2H_6 \rightleftharpoons C_2H_4 + H_2$$

2. Is reaction exothermic or endothermic?
3. Will degree of decomposition of ethane be higher than calculated in *1.* at 1 bar and 800 K?
4. Answer the same but at 2 bar and 700 K.

Suppose that the mixture is an ideal gas mixture.

Solution 1. $\triangle G^\circ = \triangle_f G^\circ{}_{C_2H_4} - \triangle_f G^\circ{}_{C_2H_6} = 34\,\text{kcal/mol} - 32\,\text{cal/mol K}\,T$

$$\triangle G^\circ_{700\,\text{K}} = 34\,\text{kcal/mol} - 32\,\frac{\text{cal}}{\text{mol K}}\,700\,\text{K} = 11\,600\,\text{cal/mol}$$

$$\prod_i a_i^{\nu_i} = \exp\left(-\frac{\triangle G^\circ}{RT}\right)$$

state	C_2H_6	$\rightleftharpoons$	C_2H_4	+	H_2	total
initial	1		0		0	1
fnal	$1-x$		x		x	$1+x$

$$\prod_i \mathfrak{a}_i^{\nu_i} = \frac{\mathfrak{a}_{C_2H_4}\mathfrak{a}_{H_2}}{\mathfrak{a}_{C_2H_6}} = \frac{\frac{x}{1+x}\frac{P}{1\,\text{bar}}\frac{x}{1+x}\frac{P}{1\,\text{bar}}}{\frac{1-x}{1+x}\frac{P}{1\,\text{bar}}} = \frac{x^2}{1-x^2}\frac{P}{1\,\text{bar}} =$$

$$= \exp\left(-\frac{11\,600\,\frac{\text{cal}}{\text{mol}}\,4.18\,\frac{\text{J}}{\text{cal}}}{8.314\,\frac{\text{cal}}{\text{mol K}}\,700\,\text{K}}\right) = 2.38\cdot 10^{-4}$$

and with $P = 1$ bar, $x = 1.5\%$.

2. From Gibbs-Helmholtz relation

$$\frac{\partial \frac{\Delta G^\circ}{T}}{\partial T} = -\frac{\Delta H^\circ}{T^2} = \frac{\partial \frac{34\,000\,\frac{\text{cal}}{\text{mol}} - 32\,\frac{\text{cal}}{\text{mol K}}T}{T}}{\partial T} = -\frac{34\,000\,\text{cal/mol}}{T^2} \Rightarrow \Delta H^\circ = 34\,000\,\frac{\text{cal}}{\text{mol}}$$

For an ideal gas mixture $\Delta H = \Delta H^\circ > 0$, is endothermic.

3 and 4. Applying $\frac{d\xi}{dT} = \frac{\Delta H}{T\Delta^2 G} > 0$, as T rises more ethane decomposes. Similarly, at constant temperature,

$$\frac{d\xi}{dP} = -\frac{\Delta V}{\Delta^2 G}$$

and for ideal gases mixtures

$$\Delta V = \sum_i \nu_i \bar{v}_i = \sum_i \nu_i \frac{RT}{P} \Rightarrow \frac{d\xi}{dP} = -\sum_i \nu_i \frac{RT}{P\Delta^2 G} < 0$$

because $\sum_i \nu_i = 1$ for the given reaction. Therefore, the equilibrium extent of reaction decreases when pressure increases.

10.8. Heterogeneous Systems

All phases must fulfll simultaneously the chemical equilibrium condition. Writing it as

$$\Delta G = \sum_{i=1}^{C} \nu_i \mu_i = 0$$

for any phase (remember that μ_i must be the same in all phases), it is an additional equilibrium condition that reduces the number of degrees of freedom of the system by one. Writing it in its practical form

$$-\frac{\Delta G^\circ}{RT} = \ln \prod_{i=1}^{C} \mathfrak{a}_i^{\nu_i} = \ln \prod_{i=1}^{C} \left(\frac{f_i}{f_i^\circ}\right)^{\nu_i}$$

as the reference state for the activity of each component i must be the standard state used for ΔG°, it must be the same for each component in all phases. Therefore, the activity of a component must be the same in all phases, so we can calculate $\mathfrak{a}_i$ in the phase where it is easier.

This allows a simplif cation in those cases where there are solid or liquid phases that consist only in one pure component, because by neglecting the effect of pressure in fugacity of a pure component solid or liquid, the activity will be 1. Naturally, care should be taken that at the equilibrium temperature neither of these components change its state from P° of the standard state to P of the system.

10.9. Several Simultaneous Reactions Systems

At constant pressure and temperature, the chemical equilibrium conditions will be the R equations

$$\forall k: \qquad \triangle G_k = \sum_{i=1}^{C} \nu_i^k \mu_i = 0 \tag{10.11}$$

where ν_i^k is the stoichiometric coeff cient of component i in reaction k.

Evidently, if any of these equations were a linear combination of others, they would be redundant and could be discarded. Therefore, we must consider only linearly independent reactions for the calculation of the equilibrium composition; this condition is fulf lled if the rank of the stoichiometric coeff cients matrix is equal to the number of reactions.

From the point of view of calculating equilibrium composition, any set of R independent reactions is equivalent, and their corresponding extent of reaction will determine the composition of the system in equilibrium from an initial known composition.

10.10. Phase Rule

For any system in equilibrium,

$$F + L = C + 2 - R - R' \tag{10.12}$$

where F is the number of phases, C is the number of components, R is the number of independent chemical reactions in the system, R' is the number of stoichiometric relations and L is the number of degrees of freedom.

The number of degrees of freedom refers here to the number of intensive variables not f xed by the equilibrium conditions. Stoichiometric relations refer to the possibility that the way in which the system operates might, as introduced in some cases, a linear relation between two (or more) variables; reducing then the number of degrees in freedom of the system.

A typical example is a closed system with just one phase in which there occurs a chemical reaction with two products, being formed initially only by reactants. In this case, the mole fractions of the product will be proportional to their stoichiometric coeff cients, no matter the extent of reaction; thus, there is an independent variable less.

It should be noted that if the system were in two phases, there would be no stoichiometric relation, because the total number of moles in each of the products would still be proportional to the stoichiometric coeff cients. This would not be the case for mole fractions in any phase; and, as it has been said, the number of degrees of freedom refer exclusively to intensive variables and not to extensive ones.

PROOF

We have seen that for an homogeneous open system with C components, the characteristic equation is:

$$U = U(S, V, n_1 \dots n_C)$$

That is, any variable, extensive or intensive, can be determined from $C+2$ adequate variables. Dividing the extensive variable by the total number of moles, specif c variables will be def ned by only $C+1$ variables, because mole fractions must always satisfy the relation

$$\sum_{i=1}^{C} x_i = 1$$

The total number of intensives variables to be determined for each phase is thus $C+1$ and in all the system $F \cdot (C+1)$.

The number of equilibrium conditions will be:

- Thermal equilibrium, $F-1$ conditions (equal temperature in every phase);
- Mechanical equilibrium, $F-1$ conditions (equal pressure in every phase, neglecting the effect of gravitational f eld and energy of the interfaces);
- Diffusion equilibrium, $(F-1) \cdot C$ conditions (equal chemical potential of each component in every phase);
- Chemical equilibrium, R conditions (partial Gibbs function of reaction equal to 0 for every reaction).

Thus we get:

$$L = F \cdot (C+1) - (F-1) - (F-1) - (F-1) \cdot C - R = C - F - R + 2$$

Taking into account the possibility of having sometimes some additional restrictions because of stoichiometric relations, we get (10.12)

Phase rule has interesting industrial applications, as for example, on designing control systems, as obviously, it is impossible to control more variables than the number of degrees in freedom of the system.

Chapter 11

Thermodynamic Industrial Processes

11.1. General

Many industrial plants are based on a set of interconnected open systems used for a specif c application. The global process that takes place in them is frequently called *thermodynamic cycle*, but it is really not cyclic in many cases.

Many applications share a same basic objective. Depending on it, industrial themodynamic 'cycles' can be classif ed as the following:

- If the purpose is to produce power, the cycle is called *thermal cycle or direct cycle*.
- If it is to cool down a particular system, the cycle is called *refrigeration cycle*.
- When the objective is to warm up a particular system, the cycle is called *heat pump*.

From the point of view of the thermodynamics of the cycle, refrigeration cycles and heat pumps are similar; their diagrams might even look the same in some cases. This is why they have a generic name: *reverse cycles*. Their main difference is their application. This def nes the working temperatures of the cycle f uid, and may require specif c topologies or f tting specif c components over the common basic arrangement. However, when referring to common aspects of heat pumps and refrigeration cycles, we shall speak of *reverse cycles*.

In this chapter we are going to analyze cyclic systems (direct and reverse) that absorb from and release heat to external systems, by an establishing an analogy with bithermal machines. For this, we shall explore their relation with Carnot cycles (see chapter 3). We shall also analyze the Brayton cycle as an example of an important internal combustion cycle. Finally, we shall consider cycles designed for liquefying gases, called 'cryogenic'.

When designing an industrial cycle, the f rst aim must be to choose a topology that will deliver the maximum performance possible and which results technologically viable. For example, when thinking of reversible bithermal machines, the ideal would be the Carnot cycle.

Logically, in real cycles the absorption and release of heat are frequently carried out in non-isothermal processes, which means a f rst difference with the Carnot ideal. However, if we considered the average temperatures at which heat is absorbed and released by the cycle and we managed to make it completely reversible in all other aspects, our cycle would have the same eff ciency as a Carnot cycle working between those temperatures.

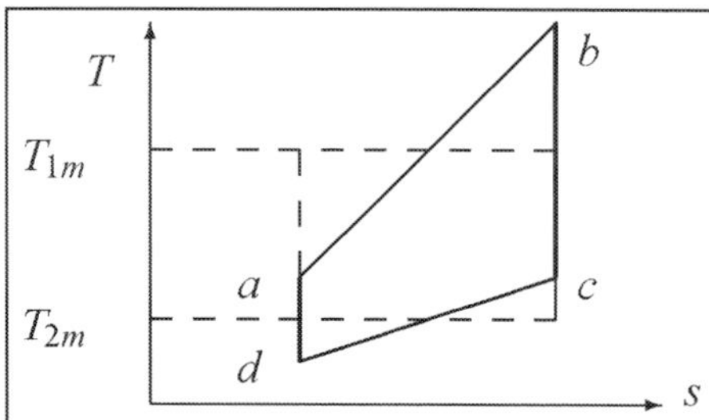

For the sake of simplicity, we consider any *internally reversible cycle abcd* in which the f uid absorbs heat in ab stage and it releases it in cd stage, considering bc and da adiabatic stages (isentropic stages). Applying the First and Second Laws:

$$0 = q_{ab} + q_{cd} - w \Rightarrow w = q_{ab} + q_{cd} \quad ; \quad q_{ab} = \int_a^b T\,ds \quad ; \quad q_{cd} = \int_c^d T\,ds$$

$$\text{Let us call} \quad T_{1m} = \frac{\int_a^b T\,ds}{s_b - s_a} \quad ; \quad T_{2m} = \frac{\int_c^d T\,ds}{s_d - s_c} \qquad (s_a = s_d \quad ; \quad s_b = s_c)$$

$$q_{ab} = T_{1m}(s_b - s_a) \quad ; \quad q_{cd} = T_{2m}(s_d - s_c) = T_{2m}(s_a - s_b)$$

$$w = (T_{1m} - T_{2m})(s_b - s_a) = \left(1 - \frac{T_{2m}}{T_{1m}}\right) q_{ab} = \frac{T_{1m} - T_{2m}}{T_{2m}}(-q_{cd})$$

In other words, both thermal performance and eff ciency of an internally reversible cycle are equal to those obtained in an equivalent Carnot cycle between the values of T_{1m} and T_{2m}. Given that these temperatures are average temperatures (average according to entropy) their values will be limited between the maximum and the minimum temperatures, the absorption and release of heat respectively.

In the following section, we shall search, parting from the Carnot cycle ideal, for a cycle design which will result technologically feasible.

11.2. Rankine Simple Cycle Direct and Reverse

The performance of a thermal Carnot cycle will improve if the temperature T_1 of the high temperature reservoir increases and T_2 of the low temperature reservoir decreases. In reality, these values are limited by:

- $T_2 = T_0$, because this low temperature reservoir is normally ambient temperature.
- $T_1 \leq T_{\max}$, where $T_{\max}$ is the maximum temperature that materials can stand (for carbon steels, it is approximately 540 °C; austenitic steels, more expensive than the previous ones, around 600 °C can be reached or even more).

Therefore, the thermal cycle with the highest performance possible is a Carnot cycle working between T_0 and $T_{\max}$. For reverse cycles (refrigeration or heat pump), eff ciency increases by decreasing the difference between T_1 and T_2. In practice, these temperatures are limited by:

- T_2: It will depend on the application of the refrigeration cycle. It will be $T_2 = T_0$ in a heat pump.

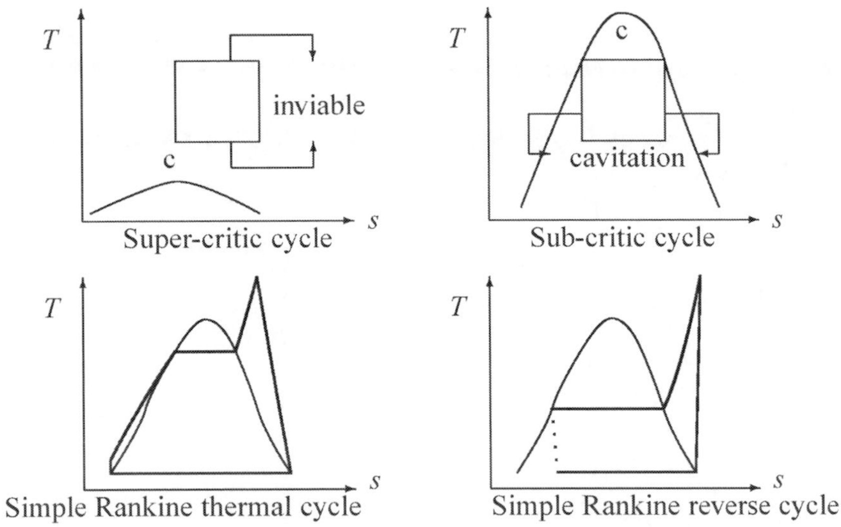

Figure 11.1.

- T_1: It will depend on the application of a heat pump. It will be $T_1 = T_0$ in a refrigeration cycle.

Either in a direct cycle or a reverse one, it is good to approach the Carnot cycle as much as possible. However, it is unattainable in practice (see f gure 11.1).

If attempting to build it in a single-phase region of the working f uid, for example supercritical, it would require to develop isothermal expansion and compression processes.

In the bi-phase region (sub-critic cycle), this inconvenient would disappear because isotherms and isobars coincide in that area. So isotherm stages would become heat exchangers. But, on the other hand, isentropic parts would be impossible to build because cavitation would appear (abrupt condensations and evaporations due to small local variations in pressure). This would reduce the components' lifetime –and performance– signif cantly.

The solution used in real designs for direct cycles consists in bringing isentropic stages out of the saturation curve. This forces to introduce new stages in order to connect the two isotherms toghether. These stages are isobaric processes at the saturation pressure corresponding to the highest isotherm. The result is a simple thermal Rankine cycle, as shown in the lower left diagram in f gure 11.1; it was already taken into account that the adiabatic processes would not be isentropic in practice, and so their lines are not vertical (there will always exist an entropy generation).

It is convenient that the turbine outlet point falls slightly inside the saturation area, because this allows a slight increase in eff ciency and power without damaging the turbine. Obviously, this cycle has a poorer thermal performance than a Carnot cycle working between the same maximum and minimum temperatures.

In reverse cycles, a different solution is used. Apart from other considerations, it would be necessary to increase the maximum pressure a lot in order to draw those sections out of the saturation curve (bear in mind that in this section, the temperature cannot be lower than the ambient temperature). Instead, a lamination is carried out in order to go from the highest pressure to the lowest pressure, as indicated by a dotted line in f gure 11.1, lower right. The irreversibility which this introduces necessarily causes a decrease in eff ciency with respect to a reversible cycle.

11.2.1. Detailed Study of the Simple Rankine Thermal Cycle

The system is formed by:

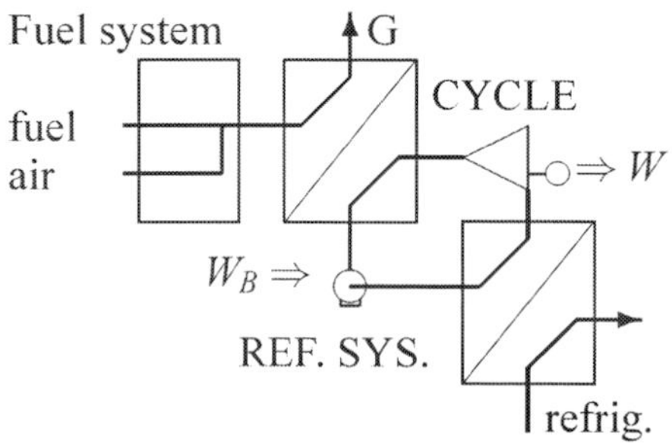

1. *Fuel System*: It supplies heat to the system that carries out the thermal cycle, obtaining energy from a chemical or nuclear combustion. The combustion is then external to the cyclic process.
2. *Closed system* that carries out the cyclic process.
3. *Refrigeration system*: Removes heat by exchanging it with a f uid whose temperature is close to the ambient temperature (water or air).

If energy and exergy balances are applied to the fuel system, i.e., to the boiler, its energy and exergy eff ciency can be expressed as the following:

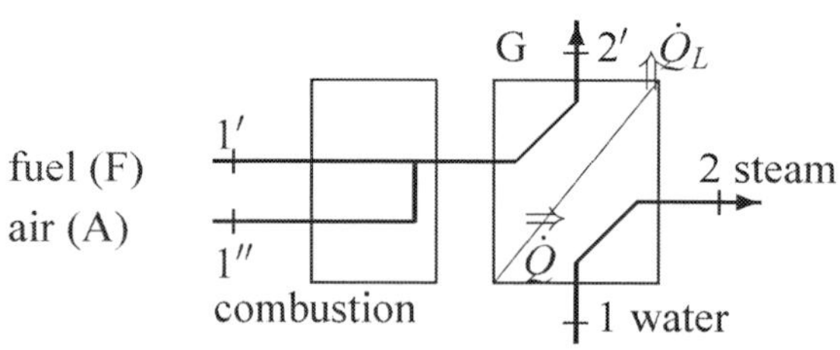

$$\eta_c = \frac{\dot{m}(h_2 - h_1)}{\dot{m}_F LHV} \tag{11.1}$$

$$\zeta_c = \frac{\dot{m}(e_2 - e_1)}{\dot{m}_F e_{1'}} \tag{11.2}$$

where $\dot{m}$ is the water inlet f ow rate in the boiler; $\dot{m}_F$, the fuel f ow rate; LHV, the lower heating value of the fuel; $\dot{e}_{1'}$, the exergy of the fuel; $\dot{e}_1$ and $\dot{e}_2$ the exergy of the inlet and outlet f ow rate of water at the boiler respectively; $\dot{h}_1$ and $\dot{h}_2$ the enthalpy of water at the inlet and outlet.

The exergy of air is obviously zero but, however, fuel exergy is not, as it has chemical exergy whose value is usually between LHV and HHV. The inlet conditions of the fuel and the air ($1'$ and $1''$ respectively) will be assumed equals to ambient ones (0). Currently, η_c can be between 0.90 and 0.92, and ζ_c, between 0.40 and 0.42.

Now we shall assess the cyclic process itself. We will neglect pressure drops due to losses in pipes and heat exchangers, the heat losses in pipes, the turbine and the pump, and the variations in kinetic and potential energy.

Applying the First and Second Laws, the thermal eff ciency of the process will be:

$$\eta = \frac{w}{q_{1-2}}$$

And the exergetic eff ciency:

$$\zeta = \frac{w}{e_2 - e_1}$$

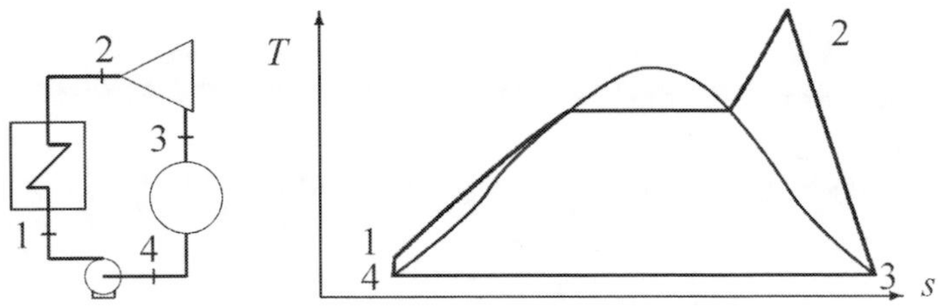

Figure 11.2.

In this expression we have not considered the exergies of the water or air f ow of the refrigeration system, assuming they are at or near ambient conditions (dead state) and that neither the heat nor the exergy could be used or recovered.

Therefore, the eff ciency of the whole system can be expressed as:

$$\eta_{\text{total}} = \frac{\dot{m}w}{\dot{m}_F LHV} = \eta_c \eta$$

In the following table, typical values for energy and exergy losses for a plant implementing a simple thermal Rankine cycle are shown:

	Energy losses	Exergy losses/destruction.
Chemical reaction	—	31%
Heat exchange	—	25.9%
Gases outf ow+heat losses	10%	4%
Turbine	—	7%
Condenser	59.2%	3.1%
Pump	—	0.1%
Total	69.2%	71.1%
Work	30.8%	28.9%

The losses for the simple thermal Rankine cycle can be classif ed as follows:

- *Structural*: cannot be reduced.
- *Operational*: can be reduced: exergy losses at the chimney, the head losses and exergy losses at the boiler (due to heat transfer between water and smoke), exergy losses by mechanical dissipation in turbines and pumps.

11.2.2. Detailed Study of the Simple Rankine Reverse Cycle

The simple reverse Rankine cycle is simpler than the direct one, because there is no fuel system. There are many possible conf gurations, so here it will be considered a reverse cycle (refrigeration) with only one release of heat to the ambient –condenser– and a heat absorption from one system at a lower temperature than the ambient –evaporator–.

By applying the First Law to the cycle, the eff ciency results:

$$\varepsilon = \frac{q_{\text{eva}}}{-w_C}$$

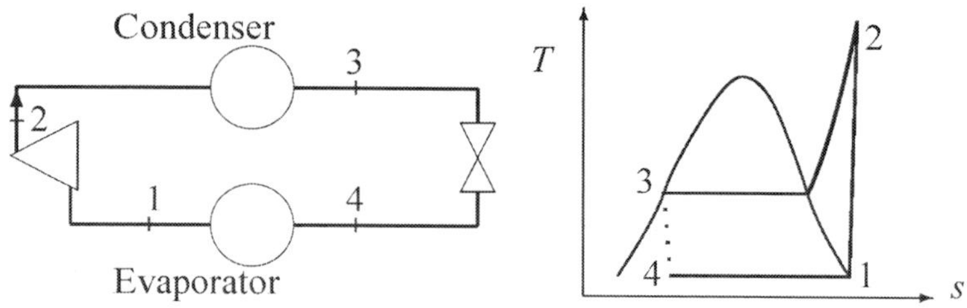

Figure 11.3. Simple Rankine reverse cycle.

And after applying the Second Principle in terms of exergy, the exergy eff ciency will be:

$$\zeta = \frac{-q_{\text{eva}}\left(1 - \frac{T_0}{T_F}\right)}{-w_C}$$

Exergy destructions referred to the exergy supply to the cycle (the duty supplied) using real data have the following orders of magnitude:

Compressor: 17.7% Lamination: 36.2%
Condenser: 20.6% Evaporator: 13.5%

11.3. Improvement Guidelines for Rankine Cycles

The performance or eff ciency of any internally reversible cycle depends exclusively on its average temperatures of release and absorption of heat. For any internally reversible cycle, the only way for improving the cycle would be changing the mentioned temperatures. Real cycles are never internally reversible; for reverse cycles there is always a lamination (and for thermal cycles occasionally as well), which is an intrinsically irrevesible process. Turbines, pumps and compressors, whose isentropic eff ciency is usually around 85% nowadays, are also irreversible.

Improvements of Rankine cycles generally involve f tting more components onto the basic topology. However, the overall result that these modif cations might have is not always clear. For example, consider that including heat exchanges in the cycle requires always a certain thermal difference between streams, which means another source of internal irreversibility added to the cycle.

It cannot be said without a more detailed analysis that the ideas which may improve the performance or eff ciency of an internally reversible cycle will necessarily lead to an improvement in the real one. Nevertheless, it is true that they do in many cases, as can be deduced from basic numerical examples.

As we have mentioned earlier, reverse cycles, whether refrigeration ones or heat pumps, are very similar thermodynamically speaking. Although the expressions for their eff ciencies differ, any improvement in one will also improve the other one, so we shall not distinguish them formally.

Therefore, the guidelines for improving Rankine cycles are:

1. Approximate to internal reversibility as much as possible, for example by using turbines or compressors with as high isentropic eff ciency as possible.

2. Increase the average warm (heat absortion) temperature of the f uid T_{1m} if the cycle is direct. Decrease it as much as possible if it is reverse.

3. Decrease the average cold (heat release) temperature of the f uid T_{2m} if the cycle is direct. Increase it is reverse.

4. Only for thermal cycles with a non-nuclear fuel: reduce the temperature of the combustion gases outlet as much as possible. The limit is often given by the appearing of corrosion problems in the chimney, due to some components condensing if the dew point is reached (depending on the case, it may be around 140 °C or higher).

In engineering terms, the previous guidelines are implemented in two ways:

- Exchange heat between two parts of the cycle (regenerative preheating or subcooling).
- Approximate turbine and compressor to isothermal behavior. For that, two stages of expansion or compression are used, with a heat exchange in the middle (middle preheating or compression in two stages).

11.3.1. Improvements for the Thermal Rankine Cycle: Regenerative Preheating and Middle Preheating

The heat exchange between two parts of the cycle can be carried out with surface exchangers (see f gure 11.4) or mixing ones (see f gure 11.5). In practice, this involves several extractions.

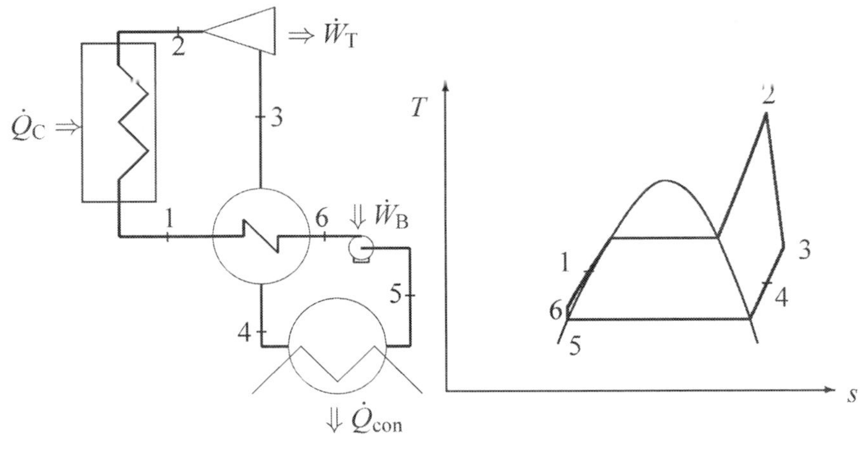

Figure 11.4. Regenerative preheating with surface exchanger.

In the example of f gure 11.4, standing for a thermal cycle, when absorbing heat the f uid in 6-1 from 3-4, the average heat absorption temperature T_{1m} is increased and the

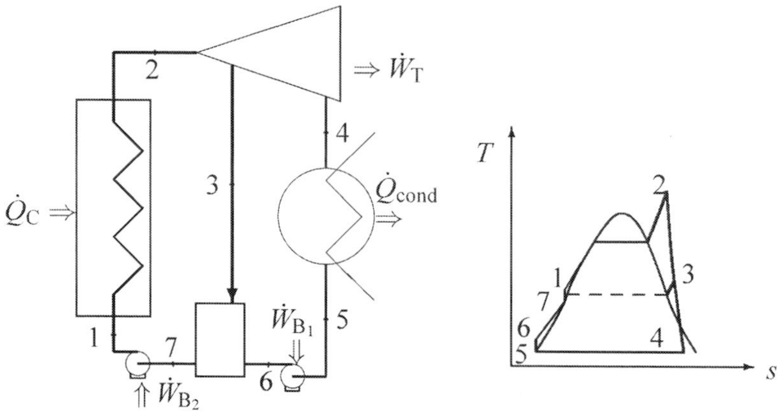

Figure 11.5. Regenerative preheating with mixing exchanger.

release temperature T_{2m} is reduced. This way of improving the performance is not used in real situations. What is done is combining it with middle preheating.

The middle preheating in the thermal cycle means a re-entrance in the boiler, that normally leads to an increase in the average heat absorption temperature. If all the steam is used, it is called total extraction, and if only part of it, partial extraction.

In the example of f gure 11.6, we can see a total extraction with preheating. As we mentioned before, this preheating increases T_{1m}. On the other hand, it increases T_{2m} as well. This effect can be corrected (increasing also T_{1m}) in combination with regenerative preheating (see the following exercise).

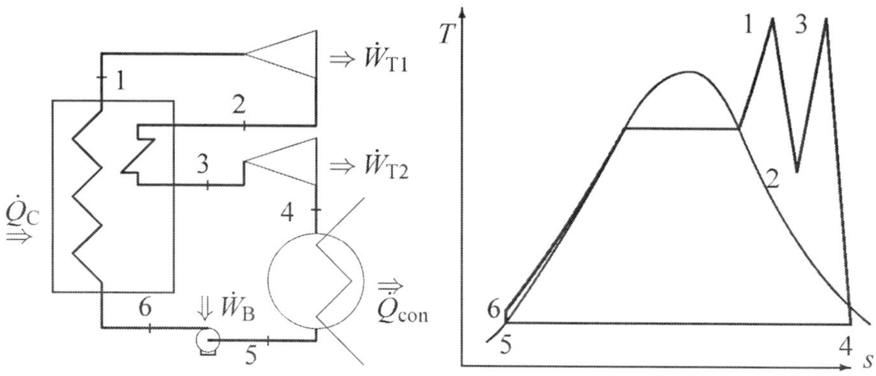

Figure 11.6. Middle preheating.

Exercise 1 Compare the thermal eff ciency of the following cycle with total extraction, middle preheating and regenerative preheating (see f gure) with that of the simple cycle between the same pressures and maximum and minimum temperatures with same isentropic eff ciencies:

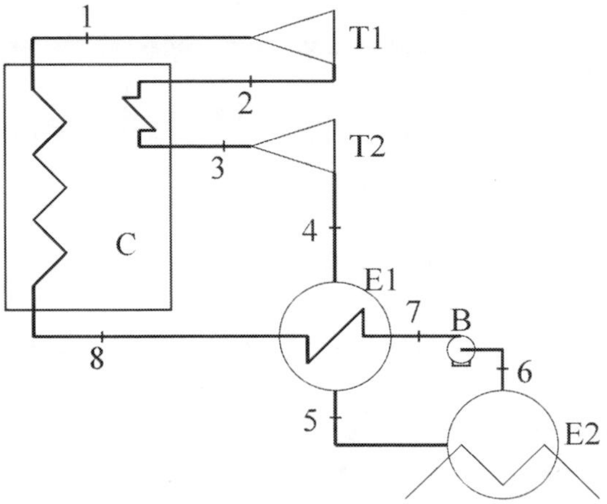

Boiler C produces steam at $P_1 = 100$ bar and $t_1 = 530\,°C$. Turbine T1 expands it to $P_2 = 3$ bar. Then it returns to the boiler, with an outlet temperature of $t_3 = 530\,°C$. It is expanded in the turbine T2 to $P_4 = 0.1$ bar. Then, it enters heat exchanger E1, from which it f ows out at $t_5 = 50\,°C$. Afterwards, it is introduced into condenser E2, from which leaves as saturated liquid. Then, it is compressed in pump B up to $P_7 = 100$ bar, it is preheated in E1 and it returns to the boiler.

Determine also exergetic destructions in each device and global exergetic eff ciency. Suppose that the exergy of the fuel is equal to its LHV. The energy eff ciency of the boiler, referred to LHV, is 0.9. Turbines and pump are adiabatic with isentropic eff ciency $\eta_s = 0.8$. Ambient temperature $t_0 = 25\,°C$. Data for water from tables: $t_6 = 45.8\,°C$; $h_6 = 191.8$ kJ/kg; $s_6 = 0.6493$ kJ/kg K. At $t_a = 60\,°C$ and $P_a = 100$ bar:$h_a = 259.4$ kJ/kg;$s_a = 0.8257$ kJ/kg K. At 100 bar,$c_p^L = 4.166$ kJ/kg K.

Solution: From Mollier diagram:
$h_1 = 3450$ kJ/kg; $s_1 = 6.695$ kJ/kg K; $h_3 = 3555$ kJ/kg; $s_3 = 8.400$ kJ/kg K
$h_5 = 2650$ kJ/kg; $s_5 = 8.345$ kJ/kg K.

$$s_{2s} = s_1 \Rightarrow h_{2s} = 2610\,\text{kJ/kg} \Rightarrow$$
$$h_2 = h_1 - \eta_s(h_1 - h_{2s}) = 2778\,\text{kJ/kg} \Rightarrow s_2 = 7.105\,\text{kJ/kg K}$$
$$s_{4s} = s_3 \Rightarrow h_{4s} = 2675\,\text{kJ/kg} \Rightarrow$$
$$h_4 = h_3 - \eta_s(h_3 - h_{4s}) = 2851\,\text{kJ/kg} \Rightarrow s_4 = 8.830\,\text{kJ/kg K}$$
$$s_{7s} = s_6 = 0.6943\,\text{kJ/kg K} = s_a + c_p^L \ln\frac{T_{7s}}{T_a} \Rightarrow$$
$$t_{7s} = 46.18\,°C \Rightarrow h_{7s} = h_a + c_p^L(t_{7s} - t_a) = 201.85\,\text{kJ/kg}$$
$$h_7 = h_6 + \frac{h_{7s} - h_6}{\eta_s} = 204.4\,\text{kJ/kg} \Rightarrow t_7 = t_a + \frac{h_7 - h_a}{c_p^L} = 46.79\,°C \Rightarrow$$
$$s_7 = s_a + c_p^L \ln\frac{T_7}{T_a} = 0.6572\,\text{kJ/kg K}$$
$$\dot{m}h_4 + \dot{m}h_7 = \dot{m}h_5 + \dot{m}h_8 \Rightarrow h_8 = 405.4\,\text{kJ/kg} = h_a + c_p^L(t_8 - t_a) \Rightarrow$$
$$t_8 = 95.05\,°C \Rightarrow s_8 = s_a + c_p^L \ln\frac{T_8}{T_a} = 1.2424\,\text{kJ/kg K}$$
$$\eta = \frac{h_1 - h_2 + h_3 - h_4 + h_6 - h_7}{h_1 - h_8 + h_3 - h_2} = 35.68\%$$

$$i_{T1} = T_0(s_2 - s_1) = 122.24\,\text{kJ/kg} \quad ; \quad i_{T2} = T_0(s_4 - s_3) = 128.20\,\text{kJ/kg}$$

$$i_B = T_0(s_7 - s_6) = 2.36\,\text{kJ/kg} \quad ; \quad i_{E1} = T_0(s_5 + s_8 - s_4 - s_7) = 29.55\,\text{kJ/kg}$$

$$i_C = \frac{h_1 - h_2 + h_3 - h_8}{0.9} - \{(h_1 - h_8) - T_0(s_1 - s_8) + (h_3 - h_2) - T_0(s_3 - s_2)\} =$$
$$= 2436.42\,\text{kJ/kg}$$

Exergy lost in condenser: $$e_L = h_5 - h_6 - T_0(s_5 - s_6) = 164.06\,\frac{\text{kJ}}{\text{kg}}$$

$$\zeta = \frac{w_{T1} + w_{T2} + w_B}{q_c/0.9} = 0.9\eta = 32.1\%$$

Bear in mind that here η is the thermal eff ciency of the cycle, fuel system excluded, and ζ the global exergy eff ciency. Summing up, the exergy destructions/losses are, according to exergy supplied by the fuel:

boiler	57.38%
turbines	5.90%
condenser	3.86%
preheating	0.70%
pump	0.06%
total	67.9%

Although the exergy destruction in the boiler due to the chemical reaction cannot be avoided (as long as Rankine cycles are used), the destruction due to heat transfer between combustion gases and water-steam can be somewhat reduced, which would increase the average warm temperature, T_{1m}.

Table 11.1. Thermodynamic values from exercise 1.

state	t (°C)	P (bar)	h (kJ/kg)	s (kJ/kg K)
1	540	130	3443.3	6.5752
2	159	5	2763.8	6.8668
3	540	5	3570.1	8.1967
4	107	0.07	2701.0	8.6486
5	50	0.07	2593.3	8.3418
6	39	0.07	163.38	0.5591
7	39	5	164.05	0.5594
8	65	5	271.73	0.8905
9	147	5	620.61	1.8139
10	149	130	636.49	1.8196

Notice that improvements for simple thermal Rankine cycle may contribute to an eff-ciency increase of around 4%. For greater improvements, it would be necessary to avoid structural losses, thus carrying out a different cycle such as the Brayton.

11.3.2. Improvements in the Reverse Rankine Cycle: Regenerative Sub-cooling and Compression in Two Stages

The heat exchanging between two parts of a reverse cycle is generically called sub-cooling. In the example of f gure 11.7, the heat exchange between 4-5 and 1-2 increases the total absorbed heat, from that of the basic cycle, which would be carried out in 6-1, to 6-2. This is done at the price of increasing the work of the compressor, but it improves the global eff ciency.

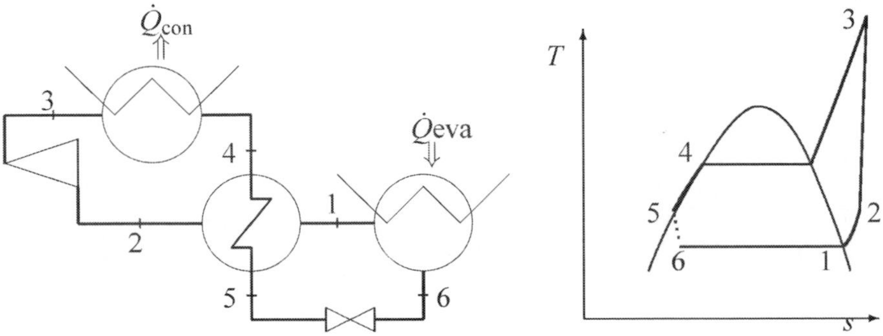

Figure 11.7. Regenerative sub-cooling.

The eff ciency of a reverse Rankine cycle improves when compression is done in two stages with intermediate heat exchange, for the same reason that two turbines with middle preheating improved the eff ciency of the direct cycle. It can be carried out with mixing exchangers (see f gure 11.9). Alternatively, with surface exchangers, as in cascade cycles (see f gure 11.8). With these exchangers, given that the streams are separated, the best f uid for each of operating conditions can be used.

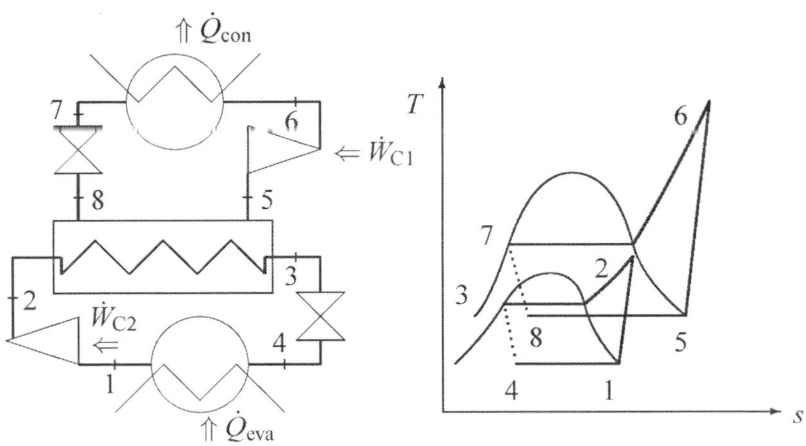

Figure 11.8. Cascade cycles.

In reverse cycles, having two compression stages generally achieves a lower difference between the average heat absorption and release temperatures, T_{2m} and T_{1m} . Moreover, the irreversibility due to lamination is also reduced, because the exergy destruction is lower in the case of two separate ones than in the case of a single one between the extreme pressures of the cycle.

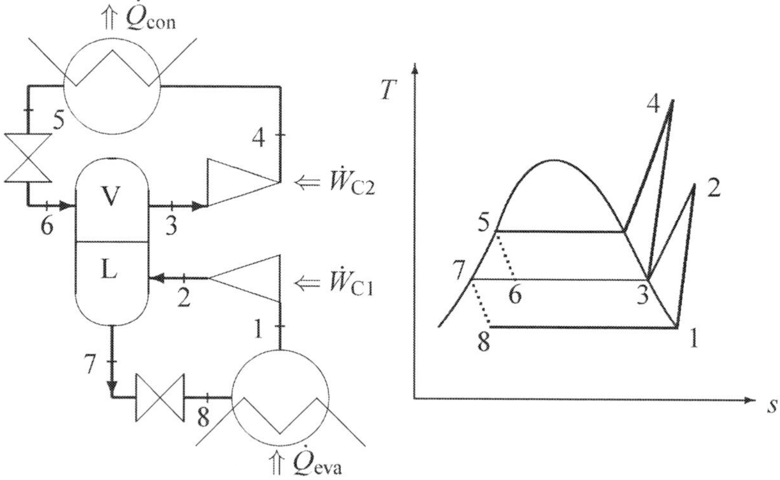

Figure 11.9. Two-stages compression with mixing exchanger.

Exercise 3: Compare the eff ciency and the global exergy performance of the double compression cycle and the simple cycle between the same pressures, with the following characteristics: isentropic eff ciency of the compressors 0.8 in both; the compressors take in saturated vapor, the heat is released to the ambient at $t_0 = 27\,°\mathrm{C}$ and it is removed from a reservoir at $t_F = -30\,°\mathrm{C}$ in both cases; the f ow rate in the evaporator is 2 t/h. Saturation curve and average specif c heats of vapor from the refrigerant f uid:

t(°C)	P (bar)	h^L (kJ/kg)	h^V (kJ/kg)	s^L (kJ/kg K)	c_p (kJ/kg K)
40	16.77	1 038.87	1 507.53	4.860 64	4.006
−4	5.05	817.96	1 443.21	4.115 30	3.042
−40	1.30	661.43	1 372.97	3.496 57	

Solution: Two-stage compression cycle (see f gure 11.9):

$$s_1 = 3.49657\,\frac{\text{kJ}}{\text{kg K}} + \frac{h_1 - 661.43\,\text{kJ/kg}}{233.15\,\text{K}} = 6.54843\,\text{kJ/kg K}$$

$$s_3 = s_7 + \frac{h_3 - h_7}{T_3} = 6.43835\,\text{kJ/kg K}$$

$$s_1 = s_{2_s} = s_3 + 3.042\,\frac{\text{kJ}}{\text{kg K}}\ln\frac{T_{2_s}}{T_3} \Rightarrow t_{2_s} = 5.92\,^\circ\text{C} \Rightarrow$$

$$h_{2_s} = h_3 + 3.042\,\frac{\text{kJ}}{\text{kg K}}(t_{2_s} - t_3) = 1\,473.38\,\text{kJ/kg}$$

$$s_3 = s_{4_s} = s_5 + \frac{1\,507.53\,\text{kJ/kg} - h_5}{T_5} + 4.006\,\frac{\text{kJ}}{\text{kg K}}\ln\frac{T_{4_s}}{T_5} \Rightarrow$$

$$t_{4_s} = 46.41\,^\circ\text{C} \Rightarrow h_{4_s} = 1\,533.19\,\text{kJ/kg}$$

$$h_2 = h_1 + \frac{h_{2_s} - h_1}{0.8} = 1\,498.48\,\text{kJ/kg} \Rightarrow t_2 = t_3 + \frac{h_2 - h_3}{3.042\,\text{kJ/kg K}} = 14.17\,^\circ\text{C} \Rightarrow$$

$$s_2 = s_3 + 3.042\,\frac{\text{kJ}}{\text{kg K}}\ln\frac{T_2}{T_3} = 6.63707\,\text{kJ/kg K}$$

$$h_4 = h_3 + \frac{h_{4_s} - h_3}{0,8} = 1\,555.68\,\text{kJ/kg} \Rightarrow t_4 = 50.02\,^\circ\text{C} \Rightarrow s_4 = 6.50813\,\text{kJ/kg K}$$

$$h_6 = h_5 = h_3 x_6 + h_7(1 - x_6) \Rightarrow x_6 = 0.35331 \Rightarrow$$

$$s_6 = s_3 x_6 + s_7(1 - x_6) = 4.93607\,\text{kJ/kg K}$$

$$h_8 = h_7 = h_1 x_8 + 661.43\,\text{kJ/kg}(1 - x_8) \Rightarrow x_8 = 0.21999 \Rightarrow$$

$$s_8 = s_1 x_8 + 3.49657\,\frac{\text{kJ}}{\text{kg K}}(1 - x_8) = 4.16794\,\text{kJ/kg K}$$

$$\dot{m}_5(h_3 - h_6) = \dot{m}_1(h_2 - h_7) \Rightarrow \dot{m}_5 = 3.366\,\text{t/h}$$

$$\dot{W}_{C1} = \dot{m}_1(h_1 - h_2) = -69.73\,\text{kW} \quad ; \quad \dot{W}_{C2} = \dot{m}_5(h_3 - h_4) = -105.16\,\text{kW}$$

$$\dot{Q}_{\text{eva}} = \dot{m}_1(h_1 - h_8) = 308.34\,\text{kW} \quad ; \quad \dot{Q}_{\text{con}} = \dot{m}_5(h_5 - h_4) = -483.22\,\text{kW}$$

$$\dot{I}_{C1} = \dot{m}_1 T_0(s_2 - s_1) = 14.78\,\text{kW} \quad ; \quad \dot{I}_{C2} = \dot{m}_5 T_0(s_4 - s_3) = 19.58\,\text{kW}$$

$$\dot{I}_{L1} = \dot{m}_1 T_0(s_8 - s_7) = 8.78\,\text{kW} \quad ; \quad \dot{I}_{L2} = \dot{m}_5 T_0(s_6 - s_5) = 21.17\,\text{kW}$$

$$\dot{I}_B = -T_0\sum_j \dot{m}_j s_j = T_0\left(\dot{m}_1(s_7 - s_2) + \dot{m}_5(s_3 - s_6)\right) = 1.1\,\text{kW}$$

$$\dot{I}_{\text{con}} = \dot{m}_5\left\{(h_4 - h_5) - T_0(s_4 - s_5) - (h_4 - h_5)\left(1 - \frac{T_0}{T_0}\right)\right\} = 20.87\,\text{kW}$$

$$\dot{I}_{\text{eva}} = \dot{m}_1(h_1 - h_8)\left(1 - \frac{T_0}{T_F}\right) - \dot{m}_1\{h_1 - h_8 - T_0(s_1 - s_8)\} = 16.33\,\text{kW}$$

$$\varepsilon = \frac{\dot{m}_1(h_1 - h_8)}{-\dot{W}_{C1} - \dot{W}_{C2}} = 1.763$$

$$\zeta = \frac{-\dot{m}_1(h_1 - h_8)\left(1 - \frac{T_0}{T_F}\right)}{-\dot{W}_{C1} - \dot{W}_{C2}} = \varepsilon\left(\frac{T_0}{T_F} - 1\right) = 41.33\%$$

Notice that $Q_{eva} > 0$ but its exergy content is negative, i.e., it is out of the system.

Simple cycle:

$$s_1 = 6.54843\,\text{kJ/kg K} = s_{2_s} = 6.35724\,\frac{\text{kJ}}{\text{kg K}} + 4.006\,\frac{\text{kJ}}{\text{kg K}}\ln\frac{T_{2_s}}{T_3} \Rightarrow$$

$$t_{2_s} = 55.31\,^\circ\text{C} \Rightarrow h_{2_s} = 1\,568.85\,\text{kJ/kg}$$

$$h_2 = h_1 + \frac{h_{2_s} - h_1}{0.8} = 1\,617.82\,\text{kJ/kg} \Rightarrow t_2 = 67.53\,^\circ\text{C} \Rightarrow s_2 = 6.69482\,\text{kJ/kg K}$$

$$\varepsilon = \frac{h_1 - h_4}{h_2 - h_1} = 1.365 \quad ; \quad \zeta = 31.99\%$$

11.3.3. Fluid Choice

When choosing the f uid for a specif c application, the following factors must be taken into account:

1. *Security*: Avoid toxic, f ammable, explosive and irritant f uids ...
2. *Investment and running cost:*
   - Low price;
   - High vaporization latent heat (decreases the size of the installation and hence its cost);
   - Saturation pressures neither very high nor very low for the considered range of temperatures (extreme pressures increase the cost of the installation)...
3. *Maintenance*: Non-soluble in lubricants, chemically inert and non-corrosive...
4. *Physical-chemical conditions*: It must not solidify between the range of conditions used, low viscosity.

Obviously, the choice will depend on each particular application. The immense majority of the thermal cycles use water as the f uid, because it complies with almost all the requirements above.

In refrigeration cycles, water would be useless because it would solidify. Nowadays, the investigation of new refrigerants has led to the substitution of the formerly widely used alkyl halides (halogenated hydrocarbons or CFC) by other f uids like hydrof uorocarbons that have no undesired environmental effects, for example HFC-134a refrigerant (CH_2F–CF_3 tetraf uoroethane) as replacement for CFC-12.

11.4. Brayton Cycle

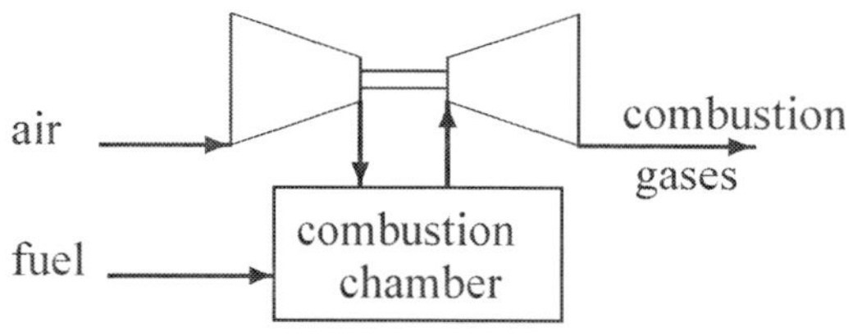

The so-called Brayton cycle is actually an open process which consists of several interconnected open systems that operate under steady state conditions and also include an internal combustion.

The gases resulting from combustion will expand in a turbine, so it is required the use of clean fuels, such as natural gas. The system, called gas turbine (see f gure) consists essentially of the following elements:

- Adiabatic compressor, which takes ambient air and compresses it up to the appropriate value.
- Combustion chamber, where compressed air and fuel are introduced. The combustion reaction can be considered complete and takes place in almost adiabatic conditions.
- Fuel system is needed to increase the pressure of the fuel from its supply network up to the one required for being introduced into the combustion chamber.
- Adiabatic turbine: drives the compressor and produces mechanical/electrical power. The inlet temperature of the turbine is limited by the materials used.

In fact, it is not possible to represent the cycle in a diagram as usual, due to the system composition changes. However, considering that the composition of the combustion gases is not too different from that of air, and that the f ow rates of air and combustion gases are quite similar, we consider the following cycle. In it, combustion (approximately adiabatic) is replaced by an isobar heat absorption.

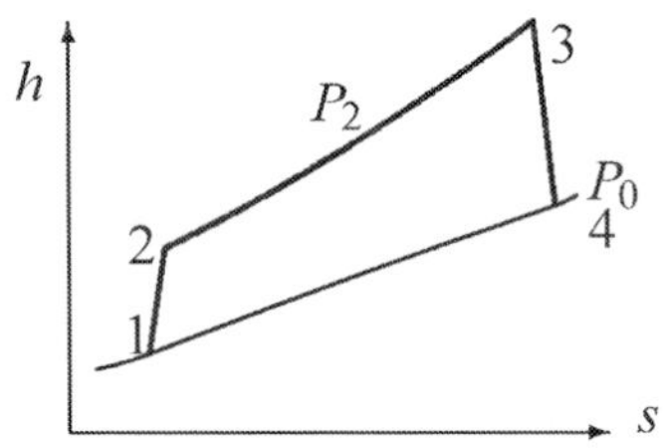

Figure 11.10. Comparative cycle for gas turbine.

The turbine exhaust gases can be used:

- Directly: for drying processes, as combustion air, and so on.
- Indirectly, in a *post-combustion* process, because the gases still have a high concentration of oxygen. Alternatively, to produce steam in a heat exchanger (*cogeneration*), frequently called *heat recovery steam generator*, HRSG, in industry.

  If the steam can be produced at high or medium pressure, it can be expanded in a steam turbine, producing a certain additional amount of work (for generating electricity, for example) and also steam (with medium or low pressure). This conf guration, which is called *combined cycle*, has a higher exergetic eff ciency.

Exercise 4: Consider the following system:

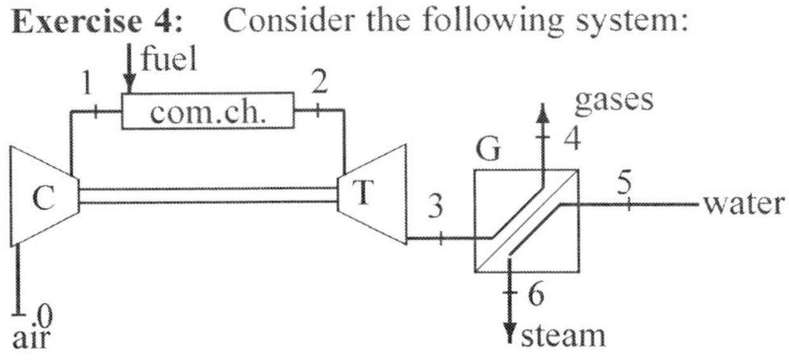

The adiabatic compressor C, with $\eta_{sc} = 0.7$ compress air taken from the ambient at $P_0 = 1$ bar and $t_0 = 20\,^\circ$C up to $P_1 = 11$ bar. This compressed air is used to f re a gaseous combustible.

The combustion gases f ow out of the combustion chamber at $P_2 = 11$ bar and $t_2 = 1000\,°\mathrm{C}$. Then, they expand in the adiabatic turbine T, with $\eta_s = 0.8$ down to $P_3 = 1$ bar. Gases coming out of T are cooled in heat exchanger G, producing 30 t/h of steam at $P_6 = 100$ bar and $t_6 = 450\,°\mathrm{C}$. The water enters in G at $P_5 = 100$ bar and $t_5 = 30.8\,°\mathrm{C}$. Calculate

1. Temperatures in 1 and 3.
2. Air mass f ow in C.
3. Power of C and T.
4. Exergetic power destroyed in C, T and G.

Air can be considered an ideal gas with $c_p^a = 1.002$ kJ/kg K, and combustion gases, as an ideal gases mixture with average specif c heat $c_p^g = 1.086$ kJ/kg K. Mass per mole both of air and combustion gases is 29 g/mol. Neglect the mass of fuel with respect to that of air. Neglect all kinetic and potential energies and heat losses. Read properties of steam from Mollier diagram. From water tables, $h_5 = 138.18$ kJ/kg, $s_5 = 0.4447$ kJ/kg K.

Solution: From Mollier diagram: $h_6 = 3245$ kJ/kg; $s_6 = 6.41$ kJ/kg K. As air and combustion gases are considered ideal gases

$$T_{1s} = T_0 \left(\frac{P_1}{P_0}\right)^{\frac{R}{c_p^a}} = 293.15\,\mathrm{K}\; 11^{\frac{8.314\frac{1}{29}}{1.002}} = 582.19\,\mathrm{K}$$

$$T_{3s} = T_2 \left(\frac{P_3}{P_2}\right)^{\frac{R}{c_p^g}} = 1273.15\,\mathrm{K} \left(\frac{1}{11}\right)^{\frac{8.314\frac{1}{29}}{1.086}} = 676\,\mathrm{K}$$

$$\eta_{sC} = \frac{h_0 - h_{1s}}{h_0 - h_1} = \frac{T_0 - T_{1s}}{T_0 - T_1} \Rightarrow T_1 = 706.6\,\mathrm{K} \equiv 432.9\,°\mathrm{C}$$

$$\eta_s = \frac{h_2 - h_3}{h_2 - h_{3s}} = \frac{T_2 - T_3}{T_2 - T_{3s}} \Rightarrow T_3 = 794.43\,\mathrm{K} \equiv 522.3\,°\mathrm{C}$$

As the fuel mass f ow is negligible compared to the air mass f ow, the mass f ow of combustion gases will be approximately equal to that of air, $\dot{m}_{\mathrm{a}}$. By the First Law at G:

$$30\,\mathrm{t/h}(h_7 - h_6) = \dot{m}_{\mathrm{a}}(h_3 - h_4) \Rightarrow \dot{m}_{\mathrm{a}} = 224.5\,\mathrm{t/h}$$

$$\dot{W}_{\mathrm{C}} = \dot{m}_{\mathrm{a}}(h_0 - h_1) = \dot{m}_{\mathrm{a}} c_p^a (t_0 - t_1) = -25796\,\mathrm{kW}$$

$$\dot{W}_{\mathrm{T}} = \dot{m}_{\mathrm{a}}(h_2 - h_3) = \dot{m}_{\mathrm{a}} c_p^g (t_2 - t_3) = 32415\,\mathrm{kW}$$

$$\dot{I}_{\mathrm{C}} = \dot{m}_{\mathrm{a}} T_0 (s_1 - s_0) = \dot{m}_{\mathrm{a}} T_0 c_p^a \ln\frac{T_1}{T_{1s}} = 3533\,\mathrm{kW}$$

$$\dot{I}_{\mathrm{T}} = \dot{m}_{\mathrm{a}} T_0 (s_3 - s_2) = \dot{m}_{\mathrm{a}} T_0 c_p^g \ln\frac{T_3}{T_{3s}} = 3204\,\mathrm{kW}$$

$$\dot{I}_{\mathrm{G}} = T_0 \sigma = -T_0 \sum_j m_j s_j = \dot{m}_{\mathrm{a}} T_0 (s_4 - s_3) + 30\,\mathrm{t/h}\; T_0 (s_7 - s_6) = 2622\,\mathrm{kW}$$

11.5. Cryogenic Cycles

The purpose of cryogenic cycles is to obtain extremely low temperatures at atmospheric pressure for liquefying gases. The most common cryogenic system is the Linde cycle. Later on, some improvements were introduced in order to increase the liquef ed f ow rate, creating the well-known Claude cycle.

In the Linde cycle, gas at ambient conditions is compressed isothermally up to a very high pressure. Afterwards, the f uid is cooled down and then it is laminated, causing partial liquefaction. In a vessel, the liquid is extracted and the gas is used for cooling down the f uid before lamination before throwing it out (see f gure 11.11).

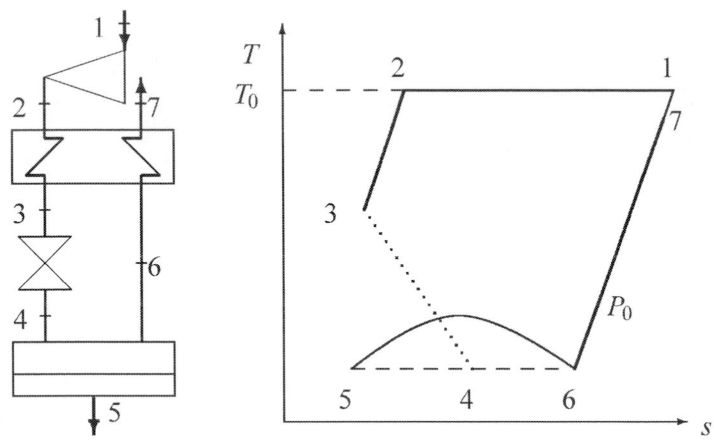

Figure 11.11. Linde cycle.

Mass and energy balances will be applied to exchanger and separator together. The relation $\frac{\dot{m}_5}{\dot{m}_2}$ is the liquid fraction obtained.

$$\dot{m}_2 = \dot{m}_5 + \dot{m}_7 \quad ; \quad \dot{m}_2 h_2 = \dot{m}_5 h_5 + \dot{m}_7 h_7 \rightarrow \frac{\dot{m}_5}{\dot{m}_2} = \frac{h_7 - h_2}{h_7 - h_5}$$

There are two important points to consider about the Linde cycle:

- Pressure at the compressor outlet that maximizes the liquid fraction. For a certain pressure, the temperature at which the Joule-Thomson coeff cient, μ_{JT} is zero is called *inversion temperature*. The geometric locus of the points at which $\mu_{JT} = 0$ is called *inversion curve*. Therefore, the largest liquid fraction is obtained when compressor outlet is at the inversion curve. In practice, it is usual to use lower pressures than this for economical reasons.
- To be able to reduce the temperature of the gas when laminating (Joule-Thomson effect), it is necessary that state 3 falls in the area where $\mu_{JT} > 0$.

Exercise 5: Consider the nitrogen liquefaction facility shown schematically in f gure 11.11. The nitrogen enters at $P_1 = 1$ bar and $T_1 = T_0 = 300$ K. It is compressed in the isothermal compressor C with $\eta_T = 0.7$ up to $P_2 = 200$ bar. Then, it goes into heat exchanger E, where it is cooled down at constant pressure and laminated in L to $P_4 = 1$ bar. After that, the liquid and vapor phases are separated in the vessel D. The outlet vapor from D is warmed in E up to $T_7 = 290$ K. For 1 t/h of liquid produced, f nd:

1. Nitrogen f ow to compress (t/h)
2. Compressor power and heat output to eliminate from the compressor.
3. Exergetic powers destroyed at L and E.

Data: Assume all elements adiabatic except C. Neglect kinetic and potential energy and pressure losses except in L. From tables of nitrogen:

$$h_1 = 1.71\,\text{kJ/kg} \qquad s_1 = 0.010\,\text{kJ/kg K}$$
$$h_2 = -30.52\,\text{kJ/kg} \qquad s_2 = -1.673\,\text{kJ/kg K}$$
$$h_5 = -431.22\,\text{kJ/kg} \qquad s_5 = -4.001\,\text{kJ/kg K}$$
$$h_6 = -232.43\,\text{kJ/kg} \qquad T_6 = 77.23\,\text{K}$$

At 1 bar and in the interval of 190 K to 350 K: $c_p^1 = 1.042$ kJ/kg K (constant).
At 200 bar and in the interval of 150 K to 300 K: $c_p^{200} = 2.706\,\text{kJ/kg K} - 0.0046\,\frac{\text{kJ}}{\text{kg K}^2}\,T$.

Solution: As 7 and 1 are on the same isobar,

$$h_7 = h_1 + c_p^1(T_7 - T_1) = -8.71\,\text{kJ/kg}$$
$$s_7 = s_1 + c_p^1 \ln\frac{T_7}{T_1} = -0.025\,\text{kJ/kg K}$$
$$h_2 = xh_5 + (1-x)h_7 \Rightarrow x = \frac{h_2 - h_7}{h_5 - h_7} = 0.051\,62$$
$$\dot{m} = \frac{1\,\text{t/h}}{x} = 19.372\,\text{t/h}$$
$$w^R = -\int_1^2 v\,dP = -(g_2 - g_1) = h_1 - h_2 - T_2(s_1 - s_2) = -472.67\,\text{kJ/kg}$$
$$\dot{W} = \dot{m}\frac{w^R}{\eta_T} = -3\,634\,\text{kW}$$
$$0 = \dot{Q} - \dot{W} + \dot{m}(h_1 - h_2) \Rightarrow \dot{Q} = -3\,807\,\text{kW}$$

$$h_2 + (1-x)h_6 = h_3 + (1-x)h_7 \Rightarrow h_3 = -242.69\,\text{kJ/kg} = h_4$$

$$h_3 = h_2 + \int_2^3 c_p^{200} dT \Rightarrow -0.0023\, T_3^2 + 2.706\, T_3 - 392.63 = 0 \Rightarrow$$

$$T_3 = \begin{cases} 1\,007\,\text{K} & \text{absurd and out of range} \\ 169.52\,\text{K} \end{cases}$$

$$s_3 = s_2 + \int_2^3 \frac{c_p^{200}}{T} dT = -2.617\,\text{kJ/kg K}$$

$$s_6 = s_5 + \frac{h_6 - h_5}{T_6} = -1.427\,\text{kJ/kg K}$$

$$s_4 = x s_5 + (1 - x) s_6 = -1.560\,\text{kJ/kg K}$$

$$\dot{I}_\text{L} = \dot{m} T_0 (s_4 - s_3) = 1\,738\,\text{kW}$$

$$\dot{I}_\text{E} = \dot{m} T_0 (s_3 - s_2) + \dot{m}(1 - x) T_0 (s_7 - s_6) = 623\,\text{kW}$$

Index